# Spon's Estimating Co
# Small Groundworks,
# and Gardening

## Also available from Spon Press

**Spon's Architects' and Builders' Price Book (2008 edition)** Hb: 978–0–415–42448–6

**Spon's External Works and Landscape Price Book (2008 edition)** Hb: 978–0–415–42444–8

**Spon's Estimating Costs Guide to Finishings: painting and decorating, plastering and tiling (2007 edition)** Pb: 978–0–415–43443–0

**Spon's Estimating Costs Guide to Plumbing & Heating (2006 edition)** Pb: 978–0–415–38618–0

**Spon's Estimating Costs Guide to Electrical Works (2006 edition)** Pb: 978–0–415–38614–2

**Spon's Estimating Costs Guide to Minor Works, Alterations and Repairs to Fire, Flood, Gale and Theft Damage (2006 edition)** Pb: 978–0–415–38213–7

**Spon's First Stage Estimating Price Book (2006 edition)** Pb: 978–0–415–38619–7

**Spon's House Improvement Price Book (2005 edition)** Pb: 978–0–415–37043–1

Information and ordering details

For price availability and ordering visit our website **www.sponpress.com**
Alternatively our books are available from all good bookshops.

# Spon's Estimating Costs Guide to Small Groundworks, Landscaping and Gardening

Second Edition

**Bryan Spain**

LONDON AND NEW YORK

First published 2005 by Taylor & Francis
Second edition published 2008 by Taylor & Francis
2 Park Square, Milton Park, Abingdon, Oxon, OX14 4RN

Simultaneously published in the USA and Canada
by Taylor & Francis
711 Third Ave, New York, NY 10017

*Taylor & Francis is an imprint of the Taylor & Francis Group, an informa business*



*Publisher's Note*
This book has been prepared from camera-ready copy supplied by the author.



*British Library Cataloguing in Publication Data*
A catalogue record for this book is available from the British Library

ISBN10: 0–415–43442–4 Paperback
ISBN13: 978–0–415–43442–3 Paperback

ISBN10: 0–203–93490–3 eBook
ISBN13: 978–0–203–93490–6 eBook

# Contents

# Preface

This is the second edition of Spon's Estimating Costs Guide to Small Groundworks, Landscaping and Gardening and is intended to provide accurate cost data for small landscaping and ground works contractors to prepare estimates and quotations more quickly and accurately.

A new section has been added dealing with groundworks covering the construction of foundations for houses and roads and sewers that would be required on small housing estates.

This need for speed and accuracy is vital for all contractors operating in today's competitive construction market. Most contractors have the skills necessary to carry out the work together with the capacity for dealing with the setbacks that are part of the normal construction process. But they rarely have enough time to complete the many tasks that must be carried out in order to trade profitably. This book aims to help contractors by providing thousands of unit rates and, if used sensibly, can save them valuable time in the preparation of their bids.

I have received a great deal of support in the research necessary for this type of book and I am grateful to those individuals and firms who have provided the cost data and other information. In particular, I am indebted to Mark Loughrey of Loughrey & Co. Ltd, Chartered Accountants of Hoylake (tel: 0151-632 3298 or mjloughrey@accounant.com), who are specialists in advising small construction businesses. Their research for the information in the business section is based on tax legislation in force in December 2006.

Although every care has been taken in the preparation of the book, neither the Publisher nor I can accept any responsibility for the use of the information provided by any firm or individual. Finally, I would welcome any constructive criticism of the book's contents and suggestions that could be incorporated into future editions.

Bryan Spain
spainandpartners@btconnect.com
April 2007

# Introduction

This edition of Spon's Estimating Costs Guide to Small Groundworks, Landscaping and Gardening follows the layout, style and contents of other books in this series. The contents of the book cover unit rates, project costs, repairs, tool and equipment hire, general advice on business matters and other information useful to those involved in the commissioning and construction of landscaping and external works. The unit rates section presents analytical rates for work up to about £50,000 in value and the business section covers advice on starting and running a business together with information on taxation and VAT matters.

## Materials

In the domestic construction market, contractors are not usually able to purchase materials in large quantities and cannot benefit from the discounts available to larger contractors. An average of 10% to 15% discount has been allowed on normal trade prices.

## Labour

The hourly labour rates for craftsmen and general operatives are based upon the current wage awards. These are set at:

| | |
|---|---|
| Craftsman | £15.00 |
| General operative | £12.00 |

These rates include provision for NIC Employers' contribution, CITB levy, insurances, public and annual holidays, severance pay and tool allowances where appropriate.

## Headings

The following column headings have been used.

| Unit | Labour | Hours £ | Materials £ | O & P £ | Total £ |
|---|---|---|---|---|---|
| m2 | 0.20 | 3.00 | 2.20 | 0.78 | 5.98 |

## Unit

This column shows the unit of measurement for the item description:

| | |
|---|---|
| nr | number |
| m | linear metre |
| m2 | square metre |
| m3 | cubic metre. |

## Labour

In the example shown, 0.20 represents the estimated time estimated to carry out one square metre of the described item, i.e. 0.20 hours.

## Hours

The entry of £3.00 is calculated by multiplying the entry in the Labour column by the labour rate of £15.00.

## Materials

This column displays the cost of the materials required to carry out one square metre of the described item, i.e. £2.20.

## O & P (Overheads and profit)

This has been set at 15% and is deemed to cover head office and site overheads including:

- heating
- lighting
- rent
- rates
- telephones
- secretarial services
- insurances
- finance charges
- transport
- small tools
- ladders
- scaffolding etc.

## Total

This is the total of the Hours, Materials and Overheads and Profit columns.

### Contracting

Tradesmen and small contractors can act as main contractors (working for a client direct) or as a subcontractor working for another contractor. Although a contract exists between a subcontractor and a main contractor, there is no contractual link between a subcontractor and an Employer.

In general terms this means that the subcontractor cannot make any claims against the Employer direct and vice-versa. It also means that the subcontractor should not accept any instructions from the Employer or his representative because this could be taken as establishing a privity of contract between the two parties,

A subcontractor must be aware of his role in the programme because if he causes a contractor to overrun the completion date for the main contract he may become liable for the full amount of liquidated damages on the main contract plus the cost of damages that the contractor and other subcontractors may have suffered.

A well-organised subcontractor will keep a full set of daily site records, staffing levels, plant on site, weather charts and such like. It also cannot be over emphasised that any verbal instructions that the subcontractor receives, should be confirmed immediately to the contractor in writing with the name of the person who issued them.

This procedure is extremely important because it may eventually save the subcontractor considerable expense if someone tries to lay the blame for delays to the contract at his door. It is also important that instructions should only be taken from the contractor and he should be informed if another party attempts to do so.

### Contractor's discount

Most sub-contracts allow for a discount to the contractor of 2½ % from the subcontractor's account. This means that the subcontractor must add this discount to his prices by adding 1/39th to his net rates.

### Payment and retention

Payment is normally made on a monthly basis. The subcontractor should submit his account to the contractor who then incorporates it into his own payment application and passes it on to the Architect or Employer's representative for certification of payment. When the subcontractor receives his payment it will be reduced by 5% retention.

This money is held by the Employer and will be released in two parts. The first part, or moiety, is paid at the completion of work and, in the subcontractor's case, this may be either when he has finished his work or when the contractor has completed the contract as a whole (known as practical completion) depending upon the contract conditions. The second part is released at the end of the defects liability period.

This money is held by the Employer and will be released in two parts. The first part, or moiety, is paid at the completion of work and, in the subcontractor's case, this may be either when he has finished his work or when the contractor has completed the contract as a whole (known as practical completion) depending upon the contract conditions. The second part is released at the end of the defects liability period.

### Defects liability period

This is the period of time (normally 12 months) during which the sub-contractor is contractually bound to return to the job to rectify any mistakes or bad pieces of workmanship. This could either be twelve months from when he completes his work or twelve months from when the main contract is completed depending upon the wording of the sub contract.

### Period for completion

Usually, a sub-contractor will be given a period of time in which he must complete the work and he must ensure that he has the capability to do the work within that period. Failure to meet the agreed completion date could have serious consequences.

Under certain circumstances, however, particularly with nominated sub-contracts, the sub-contractor may be requested to state the period of time he requires to do the work. If this is the case, then careful thought must be given to the time inserted. Too short a time may put him at financial risk but too long a time may prejudice the opportunity of winning the contract.

### Damages for non-completion

A clause is usually inserted within each sub-contract stating that the sub-contractor is liable for the financial losses that contractor suffers due to the sub-contractor's non-completion of work on time. This will include the amount of liquidated and ascertained damages contained within the main contract, together with the contractor's own direct losses and the direct losses of his other sub-contractors. As can be seen, the potential cost to the sub-contractor can be large so he must take care to expedite the work with due diligence to avoid incurring these costs.

### Variations

All sub-contracts contain a clause allowing the sub-contract work to be varied without invalidating it. The sub-contractor will normally be paid any additional cost he incurs in carrying out variations.

## Insurances

The subcontractor is responsible for insuring against injury to persons or property and against loss of plant and materials. These insurances could be taken out for each individual job, although it is more common to take out blanket policies based on the turnover the firm has achieved in the previous year.

## Extensions of time

The subcontractor will normally be entitled to a longer period of time to complete the work if he is delayed or interrupted by reasons beyond his control (known as an extension of time). Most sub-contracts list the reasons and in some cases the sub-contractor may also be entitled to additional monies as well as an extension of time.

## Domestic sub-contracts

In domestic sub-contracts the contractor would obtain competitive quotations from various subcontractors of his own choice and these may be based on a bill of quantities, specification and drawings, or schedules of work. Accompanying the enquiry should also be a form of sub-contract that the subcontractor will be required to complete.

There are several points that may affect costs and which the subcontractor should bear in mind. These are

1. Whether the rates and prices are to include for any contractor's discount (normally expressed as plus $1/39^{th}$ to allow 2½%).

2. Whether the contractor is to supply any labour or plant to assist the subcontractor in either carrying out any of the work or in off-loading materials.

3. What facilities (if any) the contractor will provide for the subcontractor such as mess rooms, welfare facilities, office accommodation and storage facilities.

4. Whether the contractor is to dispose of the subcontractor's rubbish.

## Contracting

Often a subcontractor will find himself working under a private contract, written or implied. This usually takes the form of working for a domestic householder or a small factory owner and the following procedures usually apply in this type of work.

## Estimate

The initial approach would usually come from a purchaser, e.g. 'How much will it cost to have my garden landscaped?' At this stage, he may only want an approximate cost in order to see if he can afford to have the work carried out as opposed to a quotation which is a firm offer to do the work. Therefore, a brief description of the work to be carried out together with an approximate price will suffice.

However, it should be made clear that the price is an estimate and does not constitute an offer that may be accepted by the purchaser. The estimate may be based on a telephone conversation only, e.g. 'It will cost about £4,000 to £ 5,000 to landscape your garden', or it could be based on a brief visit to the house. In either case, little time should be spent on an estimate and it is generally wise to express it as a price range.

## Quotation

A quotation is generally seen as an offer to do the work for the price quoted, and could constitute a simple contract if accepted. It follows that some time and effort should be spent in compiling a quotation to save arguments at a later stage. One should always remember that the contractor is the expert and must use his expertise in order to guide the purchaser and should discuss the work with him in full. He should tell the purchaser exactly what he is getting for the price and also what he is not.

This may mean going in to some detail such as what will happen to the surplus excavated materials, how access will be gained, how long the job will take and similar items.

The contractor should also find out from the purchaser exactly what restrictions (if any) will be placed upon him. For instance, will the purchaser keep the drive clear of cars to allow a skip to be used and will the contractor only be allowed entry to the premises on certain days and/or at certain times? These factors, should be ascertained in advance, and the costs of complying with them should be made known to the purchaser who may decide to take steps to change the restrictions.

Once the contractor has considered all the relevant factors then the formal written quotation can be produced. It should state precisely what the purchaser is getting for his money, including when and how long the job will take and contain all the salient points of discussions that have taken place.

After a quotation has been submitted then all that needs to be done is for the purchaser to accept it. Although a verbal acceptance would constitute a binding agreement, it is always more satisfactory if the acceptance is made in writing.

## Payments

There is much debate on how and when payments should be made in domestic situations. Ideally from a contractor's point of view to be paid in advance would be the most advantageous, but the chances of the purchaser wishing to do this are remote.

On the other hand, it may cause undue financial hardship to a recently self-employed contractor to have to buy all the materials himself and not get paid until all the work is completed. Whatever payment policy is adopted it must be agreed with the purchaser in advance and form part of the written quotation.

Possible alternatives are

1. Being paid when the work is complete. This is probably the best method from a public relations aspect and contractors who can complete a job in a few days should have no difficulty in adopting this policy.

2. Being paid before the work is done. This is only really feasible where the contractor concerned is of unquestionable reputation or is well known to the purchaser.

3. Being paid for materials as they are bought and delivered with the balance paid when the work is complete. This could be a practical solution for smaller contractors, but the purchaser will probably want proof of the material costs, so careful handling of invoices is necessary.

4. Some form of stage payments that usually take the form of agreed percentages of the quotation price or agreed parts of the quotation price paid after stages of the work have been carried out.

## Pricing and variations

It is important that some method of recording, pricing and being paid for variations is agreed at the outset and this is particularly relevant when dealing with private clients. Unforeseen additions, more than any other item, are the main cause of disputes and are often avoidable.

The risk of this type of dispute can be reduced by ensuring that the original quotation is as detailed as possible. The detailed specification of the materials could be contained within the descriptions or done separately. A quotation broken down in this way is detailed enough to enable the purchaser to ascertain that he is not being overcharged for any variations that may occur and yet is not so detailed that the purchaser is going to question the price of every detail.

Also, if the purchaser should wish to change anything himself then there are no arguments on what was included in the original quotation.
If variations occur, it must be established who should pay for them. There are three main types of variations.

1. Those instructed by the purchaser.

2. Those that should have been included in the original quotation.

3. Those that are necessary due to events that could not have been foreseen.

The liabilities for 1 and 2 are relatively straightforward. If the purchaser says he wants a different paving flag to his original choice, then he must bear the additional cost. Conversely, if the contractor forgot to include the cost of the sub-base in his quotation then it is only fair that he bears the cost.

Item 3 is more difficult. If it is the purchaser who is receiving the benefit of the variations and if they were not foreseeable, then it would be logical to assume that it is the purchaser who should bear the cost. An example would be where the excavation to a patio revealed old foundations underneath, the contractor would expect to be paid the extra cost for removing them.

Other instances may not be as clear cut as this example and it may become necessary to arrive at a cost-sharing arrangement if genuine doubt exists. Variations should preferably be agreed in advance before the work is carried out. They should be recorded and signed by both parties and, wherever possible, priced in detail and agreed.

# Standard Method of Measurement/trades links

The contents of this book are presented under trade headings and the following table provides a link to the Standard Method of Measurement (SMM7).

**External works**

R12 Drainage below ground
Q40 Fencing
Q10 Concrete kerbs and edgings
Q20 Hardcore/granular sub-bases
Q25 Slab/brick pavings

# Part One

## UNIT RATES

Soft landscaping

- Soil stabilisation
- Excavation and filling
- Seeding and turfing
- Bare root trees
- Container grown trees
- Conifers
- Shrubs
- Climbers
- Herbaceous plants
- Hedging
- Bedding plants
- Maintenance
- Sundries

| | Unit | Labour | Hours £ | Mat'ls £ | O & P £ | Total £ |
|---|---|---|---|---|---|---|
| **SOIL STABILISATION** | | | | | | |
| Biogradable unseeded erosion control mats 2400mm wide fixed with pins to prepared ground | | | | | | |
| Eromat Light | m2 | 0.05 | 0.75 | 2.56 | 0.50 | 3.81 |
| Eromat Standard | m2 | 0.05 | 0.75 | 2.63 | 0.51 | 3.89 |
| Eromat Coco | m2 | 0.05 | 0.75 | 2.65 | 0.51 | 3.91 |
| Biogradable seeded erosion control mats 2400mm wide fixed with pins to prepared ground | | | | | | |
| Covamat Standard | m2 | 0.05 | 0.75 | 2.92 | 0.55 | 4.22 |
| Covamat Special | m2 | 0.05 | 0.75 | 3.14 | 0.58 | 4.47 |
| Covamat Coco | m2 | 0.05 | 0.75 | 3.14 | 0.58 | 4.47 |
| **EXCAVATION AND FILLING** | | | | | | |
| **Excavation by hand** | | | | | | |
| The following rates are based on excavating in firm ground. The following adjustments should be made for other conditions: | | | | | | |
| stiff clay + 50%<br>soft chalk + 100% | | | | | | |
| Remove undergrowth and site vegetation | m2 | 0.15 | 2.25 | - | 0.34 | 2.59 |

| | Unit | Labour | Hours £ | Mat'ls £ | O & P £ | Total £ |
|---|---|---|---|---|---|---|
| Cut down trees, grub up roots and remove | | | | | | |
| girth, 600–1500mm | nr | 20.00 | 300.00 | - | 45.00 | 345.00 |
| girth, 1500–3000mm | nr | 44.00 | 660.00 | - | 99.00 | 759.00 |
| Cut down hedge, grub up roots and remove | | | | | | |
| height, 1500mm | m | 2.00 | 30.00 | - | 4.50 | 34.50 |
| height, 3000mm | m | 2.80 | 42.00 | - | 6.30 | 48.30 |
| Excavate topsoil or turf and lay aside for re-use | | | | | | |
| 150mm thick | m2 | 0.35 | 5.25 | - | 0.79 | 6.04 |
| 200mm thick | m2 | 0.45 | 6.75 | - | 1.01 | 7.76 |
| 250mm thick | m2 | 0.60 | 9.00 | - | 1.35 | 10.35 |
| Excavate to reduce levels depth not exceeding | | | | | | |
| 250mm thick | m3 | 2.20 | 33.00 | - | 4.95 | 37.95 |
| 500mm thick | m3 | 2.40 | 36.00 | - | 5.40 | 41.40 |
| Excavate trenches depth not exceeding | | | | | | |
| 250mm thick | m3 | 2.40 | 36.00 | - | 5.40 | 41.40 |
| 500mm thick | m3 | 2.50 | 37.50 | - | 5.63 | 43.13 |
| 1000mm thick | m3 | 2.60 | 39.00 | - | 5.85 | 44.85 |
| 1500mm thick | m3 | 2.80 | 42.00 | - | 6.30 | 48.30 |

| | Unit | Labour | Hours £ | Mat'ls £ | O & P £ | Total £ |
|---|---|---|---|---|---|---|
| Excavate pits depth not exceeding | | | | | | |
| 250mm thick | m3 | 2.60 | 39.00 | - | 5.85 | 44.85 |
| 500mm thick | m3 | 2.70 | 40.50 | - | 6.08 | 46.58 |
| 1000mm thick | m3 | 2.80 | 42.00 | - | 6.30 | 48.30 |
| 1500mm thick | m3 | 2.90 | 43.50 | - | 6.53 | 50.03 |
| Extra for excavating through | | | | | | |
| rock | m3 | 10.00 | 150.00 | - | 22.50 | 172.50 |
| concrete | m3 | 8.00 | 120.00 | - | 18.00 | 138.00 |
| brickwork | m3 | 6.00 | 90.00 | - | 13.50 | 103.50 |
| **Disposal by hand** | | | | | | |
| Load surplus excavated material into barrows, wheel and deposit in temporary spoil heaps, skip or lorry | | | | | | |
| distance, 25m | m3 | 1.20 | 18.00 | - | 2.70 | 20.70 |
| distance, 50m | m3 | 2.00 | 30.00 | - | 4.50 | 34.50 |
| Load surplus excavated material into barrows, wheel and spread and level on site | | | | | | |
| distance, 25m | m3 | 1.40 | 21.00 | - | 3.15 | 24.15 |
| distance, 50m | m3 | 2.20 | 33.00 | - | 4.95 | 37.95 |

| | Unit | Labour | Hours £ | Mat'ls £ | O & P £ | Total £ |
|---|---|---|---|---|---|---|
| **Filling by hand** | | | | | | |
| Surplus excavated material deposited and compacted in layers | | | | | | |
| over 250mm thick | m3 | 1.20 | 18.00 | - | 2.70 | 20.70 |
| 100mm thick | m2 | 0.20 | 3.00 | - | 0.45 | 3.45 |
| 150mm thick | m2 | 0.35 | 5.25 | - | 0.79 | 6.04 |
| 200mm thick | m2 | 0.50 | 7.50 | - | 1.13 | 8.63 |
| Imported sand deposited and compacted in layers | | | | | | |
| over 250mm thick | m3 | 1.20 | 18.00 | 33.91 | 7.79 | 59.70 |
| 100mm thick | m2 | 0.20 | 3.00 | 3.91 | 1.04 | 7.95 |
| 150mm thick | m2 | 0.35 | 5.25 | 5.86 | 1.67 | 12.78 |
| 200mm thick | m2 | 0.50 | 7.50 | 7.82 | 2.30 | 17.62 |
| Imported hardcore deposited and compacted in layers | | | | | | |
| over 250mm thick | m3 | 1.20 | 18.00 | 20.02 | 5.70 | 43.72 |
| 100mm thick | m2 | 0.20 | 3.00 | 2.02 | 0.75 | 5.77 |
| 150mm thick | m2 | 0.35 | 5.25 | 3.03 | 1.24 | 9.52 |
| 200mm thick | m2 | 0.50 | 7.50 | 4.04 | 1.73 | 13.27 |
| Imported topsoil deposited and compacted in layers | | | | | | |
| over 250mm thick | m3 | 1.20 | 18.00 | 15.20 | 4.98 | 38.18 |
| 100mm thick | m2 | 0.20 | 3.00 | 1.52 | 0.68 | 5.20 |
| 150mm thick | m2 | 0.35 | 5.25 | 2.28 | 1.13 | 8.66 |
| 200mm thick | m2 | 0.50 | 7.50 | 3.04 | 1.58 | 12.12 |

| | Unit | Plant £ | Mat'ls £ | O & P £ | Total £ |
|---|---|---|---|---|---|
| **Excavation by machine** | | | | | |
| Where applicable the plant column includes the cost of the operator. | | | | | |
| The following rates are based on excavating in firm ground. The following adjustments should be made for other conditions: | | | | | |
| stiff clay + 50%<br>soft chalk + 100% | | | | | |
| Remove undergrowth and site vegetation | m2 | 0.17 | - | 0.03 | 0.20 |
| Cut down trees, grub up roots and remove | | | | | |
| girth, 600-1500mm | nr | 200.00 | - | 30.00 | 230.00 |
| girth, 1500-3000mm | nr | 527.60 | - | 79.14 | 606.74 |
| Cut down hedge, grub up roots and remove | | | | | |
| height, 1500mm | m | 17.36 | - | 2.60 | 19.96 |
| height, 3000mm | m | 30.40 | - | 4.56 | 34.96 |
| Excavate topsoil or turf and lay aside for re-use | | | | | |
| 150mm thick | m2 | 0.33 | - | 0.05 | 0.38 |
| 200mm thick | m2 | 0.44 | - | 0.07 | 0.51 |
| 250mm thick | m2 | 0.54 | - | 0.08 | 0.62 |

| | Unit | Plant £ | Mat'ls £ | O & P £ | Total £ |
|---|---|---|---|---|---|
| Excavate to reduce levels depth not exceeding | | | | | |
| 250mm thick | m3 | 1.62 | - | 0.24 | 1.86 |
| 500mm thick | m3 | 1.52 | - | 0.23 | 1.75 |
| Excavate trenches depth not exceeding | | | | | |
| 250mm thick | m3 | 3.47 | - | 0.52 | 3.99 |
| 500mm thick | m3 | 3.37 | - | 0.51 | 3.88 |
| 1000mm thick | m3 | 3.26 | - | 0.49 | 3.75 |
| 1500mm thick | m3 | 3.15 | - | 0.47 | 3.62 |
| Excavate pits depth not exceeding | | | | | |
| 250mm thick | m3 | 3.58 | - | 0.54 | 4.12 |
| 500mm thick | m3 | 3.47 | - | 0.52 | 3.99 |
| 1000mm thick | m3 | 3.37 | - | 0.51 | 3.88 |
| 1500mm thick | m3 | 3.26 | - | 0.49 | 3.75 |
| Extra for excavating through | | | | | |
| rock | m3 | 71.65 | - | 10.75 | 82.40 |
| concrete | m3 | 60.80 | - | 9.12 | 69.92 |
| brickwork | m3 | 47.76 | - | 7.16 | 54.92 |
| **Disposal by machine** | | | | | |
| Load surplus excavated material into lorries and cart away to tip | | | | | |
| distance, 10km | m3 | 17.96 | - | 2.69 | 20.65 |
| distance, 15km | m3 | 19.54 | - | 2.93 | 22.47 |

| | Unit | Plant £ | Mat'ls £ | O & P £ | Total £ |
|---|---|---|---|---|---|
| **Filling by machine** | | | | | |
| Surplus excavated material deposited and compacted in layers | | | | | |
| over 250mm thick | m3 | 7.84 | - | 1.18 | 9.02 |
| 100mm thick | m2 | 1.20 | - | 0.18 | 1.38 |
| 150mm thick | m2 | 1.35 | - | 0.20 | 1.55 |
| 200mm thick | m2 | 1.50 | - | 0.23 | 1.73 |
| Imported sand deposited and compacted in layers | | | | | |
| over 250mm thick | m3 | 7.84 | 33.91 | 6.26 | 48.01 |
| 100mm thick | m2 | 1.20 | 3.39 | 0.69 | 5.28 |
| 150mm thick | m2 | 1.35 | 5.08 | 0.96 | 7.39 |
| 200mm thick | m2 | 1.50 | 7.78 | 1.39 | 10.67 |
| Imported hardcore deposited and compacted in layers | | | | | |
| over 250mm thick | m3 | 7.84 | 20.02 | 4.18 | 32.04 |
| 100mm thick | m2 | 1.20 | 2.02 | 0.48 | 3.70 |
| 150mm thick | m2 | 1.35 | 3.03 | 0.66 | 5.04 |
| 200mm thick | m2 | 1.50 | 4.04 | 0.83 | 6.37 |
| Imported topsoil deposited and compacted in layers | | | | | |
| over 250mm thick | m3 | 7.84 | 15.20 | 3.46 | 26.50 |
| 100mm thick | m2 | 1.20 | 1.52 | 0.41 | 3.13 |
| 150mm thick | m2 | 1.35 | 2.28 | 0.54 | 4.17 |
| 200mm thick | m2 | 1.50 | 3.04 | 0.68 | 5.22 |

| | Unit | Labour | Hours £ | Mat'ls £ | O & P £ | Total £ |
|---|---|---|---|---|---|---|
| **SEEDING AND TURFING** | | | | | | |
| Imported topsoil deposited in spoil heaps | m3 | - | - | 15.22 | 2.28 | 17.50 |
| **Pre-seeding work by hand** | | | | | | |
| Lift topsoil from spoil heaps and spread and level in layers | | | | | | |
| 75mm thick | m2 | 0.03 | 0.45 | - | 0.07 | 0.52 |
| 100mm thick | m2 | 0.05 | 0.75 | - | 0.11 | 0.86 |
| 150mm thick | m2 | 0.07 | 1.05 | - | 0.16 | 1.21 |
| Rake topsoil to a fine tilth | m2 | 0.04 | 0.60 | - | 0.09 | 0.69 |
| **Grass seeding by hand** | | | | | | |
| Sow grass seed on prepared ground | | | | | | |
| PC £60.00 per 25kg | | | | | | |
| 10g per m2 | m2 | 0.01 | 0.15 | 0.03 | 0.03 | 0.21 |
| 12g per m2 | m2 | 0.01 | 0.15 | 0.03 | 0.03 | 0.21 |
| 14g per m2 | m2 | 0.01 | 0.15 | 0.04 | 0.03 | 0.22 |
| 16g per m2 | m2 | 0.01 | 0.15 | 0.04 | 0.03 | 0.22 |
| 18g per m2 | m2 | 0.01 | 0.15 | 0.04 | 0.03 | 0.22 |
| 20g per m2 | m2 | 0.01 | 0.15 | 0.05 | 0.03 | 0.23 |
| 22g per m2 | m2 | 0.01 | 0.15 | 0.05 | 0.03 | 0.23 |
| 24g per m2 | m2 | 0.01 | 0.15 | 0.05 | 0.03 | 0.23 |
| 26g per m2 | m2 | 0.02 | 0.30 | 0.06 | 0.05 | 0.41 |
| 28g per m2 | m2 | 0.02 | 0.30 | 0.06 | 0.05 | 0.41 |
| 30g per m2 | m2 | 0.02 | 0.30 | 0.07 | 0.06 | 0.43 |

| | Unit | Labour | Hours £ | Mat'ls £ | O & P £ | Total £ |
|---|---|---|---|---|---|---|
| 32g per m2 | m2 | 0.02 | 0.30 | 0.07 | 0.06 | 0.43 |
| 34g per m2 | m2 | 0.02 | 0.30 | 0.07 | 0.06 | 0.43 |
| 36g per m2 | m2 | 0.02 | 0.30 | 0.08 | 0.06 | 0.44 |
| 38g per m2 | m2 | 0.02 | 0.30 | 0.08 | 0.06 | 0.44 |
| 40g per m2 | m2 | 0.03 | 0.45 | 0.09 | 0.08 | 0.62 |
| 42g per m2 | m2 | 0.03 | 0.45 | 0.09 | 0.08 | 0.62 |
| 44g per m2 | m2 | 0.03 | 0.45 | 0.09 | 0.08 | 0.62 |
| 46g per m2 | m2 | 0.03 | 0.45 | 0.10 | 0.08 | 0.63 |
| 48g per m2 | m2 | 0.03 | 0.45 | 0.10 | 0.08 | 0.63 |
| 50g per m2 | m2 | 0.03 | 0.45 | 0.10 | 0.08 | 0.63 |
| PC £70.00 per 25kg | | | | | | |
| 10g per m2 | m2 | 0.01 | 0.15 | 0.04 | 0.03 | 0.22 |
| 12g per m2 | m2 | 0.01 | 0.15 | 0.04 | 0.03 | 0.22 |
| 14g per m2 | m2 | 0.01 | 0.15 | 0.05 | 0.03 | 0.23 |
| 16g per m2 | m2 | 0.01 | 0.15 | 0.05 | 0.03 | 0.23 |
| 18g per m2 | m2 | 0.01 | 0.15 | 0.06 | 0.03 | 0.24 |
| 20g per m2 | m2 | 0.01 | 0.15 | 0.07 | 0.03 | 0.25 |
| 22g per m2 | m2 | 0.01 | 0.15 | 0.07 | 0.03 | 0.25 |
| 24g per m2 | m2 | 0.01 | 0.15 | 0.08 | 0.03 | 0.26 |
| 26g per m2 | m2 | 0.02 | 0.30 | 0.09 | 0.06 | 0.45 |
| 28g per m2 | m2 | 0.02 | 0.30 | 0.09 | 0.06 | 0.45 |
| 30g per m2 | m2 | 0.02 | 0.30 | 0.10 | 0.06 | 0.46 |
| 32g per m2 | m2 | 0.02 | 0.30 | 0.10 | 0.06 | 0.46 |
| 34g per m2 | m2 | 0.02 | 0.30 | 0.11 | 0.06 | 0.47 |
| 36g per m2 | m2 | 0.02 | 0.30 | 0.12 | 0.06 | 0.48 |
| 38g per m2 | m2 | 0.02 | 0.30 | 0.12 | 0.06 | 0.48 |
| 40g per m2 | m2 | 0.03 | 0.45 | 0.13 | 0.09 | 0.67 |
| 42g per m2 | m2 | 0.03 | 0.45 | 0.13 | 0.09 | 0.67 |
| 44g per m2 | m2 | 0.03 | 0.45 | 0.14 | 0.09 | 0.68 |
| 46g per m2 | m2 | 0.03 | 0.45 | 0.14 | 0.09 | 0.68 |
| 48g per m2 | m2 | 0.03 | 0.45 | 0.15 | 0.09 | 0.69 |
| 50g per m2 | m2 | 0.03 | 0.45 | 0.16 | 0.09 | 0.70 |

| | Unit | Labour | Hours £ | Mat'ls £ | O & P £ | Total £ |
|---|---|---|---|---|---|---|
| **Grass seeding by hand (cont'd)** | | | | | | |
| PC £80.00 per 25kg | | | | | | |
| 10g per m2 | m2 | 0.01 | 0.15 | 0.04 | 0.03 | 0.22 |
| 12g per m2 | m2 | 0.01 | 0.15 | 0.04 | 0.03 | 0.22 |
| 14g per m2 | m2 | 0.01 | 0.15 | 0.05 | 0.03 | 0.23 |
| 16g per m2 | m2 | 0.01 | 0.15 | 0.05 | 0.03 | 0.23 |
| 18g per m2 | m2 | 0.01 | 0.15 | 0.06 | 0.03 | 0.24 |
| 20g per m2 | m2 | 0.01 | 0.15 | 0.07 | 0.03 | 0.25 |
| 22g per m2 | m2 | 0.01 | 0.15 | 0.07 | 0.03 | 0.25 |
| 24g per m2 | m2 | 0.01 | 0.15 | 0.08 | 0.03 | 0.26 |
| 26g per m2 | m2 | 0.02 | 0.30 | 0.09 | 0.06 | 0.45 |
| 28g per m2 | m2 | 0.02 | 0.30 | 0.09 | 0.06 | 0.45 |
| 30g per m2 | m2 | 0.02 | 0.30 | 0.10 | 0.06 | 0.46 |
| 32g per m2 | m2 | 0.02 | 0.30 | 0.10 | 0.06 | 0.46 |
| 34g per m2 | m2 | 0.02 | 0.30 | 0.11 | 0.06 | 0.47 |
| 36g per m2 | m2 | 0.02 | 0.30 | 0.12 | 0.06 | 0.48 |
| 38g per m2 | m2 | 0.02 | 0.30 | 0.12 | 0.06 | 0.48 |
| 40g per m2 | m2 | 0.03 | 0.45 | 0.13 | 0.09 | 0.67 |
| 42g per m2 | m2 | 0.03 | 0.45 | 0.13 | 0.09 | 0.67 |
| 44g per m2 | m2 | 0.03 | 0.45 | 0.14 | 0.09 | 0.68 |
| 46g per m2 | m2 | 0.03 | 0.45 | 0.14 | 0.09 | 0.68 |
| 48g per m2 | m2 | 0.03 | 0.45 | 0.15 | 0.09 | 0.69 |
| 50g per m2 | m2 | 0.03 | 0.45 | 0.16 | 0.09 | 0.70 |
| PC £90.00 per 25kg | | | | | | |
| 10g per m2 | m2 | 0.01 | 0.15 | 0.04 | 0.03 | 0.22 |
| 12g per m2 | m2 | 0.01 | 0.15 | 0.04 | 0.03 | 0.22 |
| 14g per m2 | m2 | 0.01 | 0.15 | 0.05 | 0.03 | 0.23 |
| 16g per m2 | m2 | 0.01 | 0.15 | 0.05 | 0.03 | 0.23 |
| 18g per m2 | m2 | 0.01 | 0.15 | 0.06 | 0.03 | 0.24 |
| 20g per m2 | m2 | 0.01 | 0.15 | 0.07 | 0.03 | 0.25 |
| 22g per m2 | m2 | 0.01 | 0.15 | 0.07 | 0.03 | 0.25 |
| 24g per m2 | m2 | 0.01 | 0.15 | 0.08 | 0.03 | 0.26 |
| 26g per m2 | m2 | 0.02 | 0.30 | 0.09 | 0.06 | 0.45 |

| | Unit | Labour | Hours £ | Mat'ls £ | O & P £ | Total £ |
|---|---|---|---|---|---|---|
| 28g per m2 | m2 | 0.02 | 0.30 | 0.09 | 0.06 | 0.45 |
| 30g per m2 | m2 | 0.02 | 0.30 | 0.10 | 0.06 | 0.46 |
| 32g per m2 | m2 | 0.02 | 0.30 | 0.10 | 0.06 | 0.46 |
| 34g per m2 | m2 | 0.02 | 0.30 | 0.11 | 0.06 | 0.47 |
| 36g per m2 | m2 | 0.02 | 0.30 | 0.12 | 0.06 | 0.48 |
| 38g per m2 | m2 | 0.02 | 0.30 | 0.12 | 0.06 | 0.48 |
| 40g per m2 | m2 | 0.03 | 0.45 | 0.13 | 0.09 | 0.67 |
| 42g per m2 | m2 | 0.03 | 0.45 | 0.13 | 0.09 | 0.67 |
| 44g per m2 | m2 | 0.03 | 0.45 | 0.14 | 0.09 | 0.68 |
| 46g per m2 | m2 | 0.03 | 0.45 | 0.14 | 0.09 | 0.68 |
| 48g per m2 | m2 | 0.03 | 0.45 | 0.15 | 0.09 | 0.69 |
| 50g per m2 | m2 | 0.03 | 0.45 | 0.16 | 0.09 | 0.70 |
| PC £100.00 per 25kg | | | | | | |
| 10g per m2 | m2 | 0.01 | 0.15 | 0.05 | 0.03 | 0.23 |
| 12g per m2 | m2 | 0.01 | 0.15 | 0.06 | 0.03 | 0.24 |
| 14g per m2 | m2 | 0.01 | 0.15 | 0.07 | 0.03 | 0.25 |
| 16g per m2 | m2 | 0.01 | 0.15 | 0.07 | 0.03 | 0.25 |
| 18g per m2 | m2 | 0.01 | 0.15 | 0.08 | 0.03 | 0.26 |
| 20g per m2 | m2 | 0.01 | 0.15 | 0.09 | 0.04 | 0.28 |
| 22g per m2 | m2 | 0.01 | 0.15 | 0.10 | 0.04 | 0.29 |
| 24g per m2 | m2 | 0.01 | 0.15 | 0.11 | 0.04 | 0.30 |
| 26g per m2 | m2 | 0.02 | 0.30 | 0.11 | 0.06 | 0.47 |
| 28g per m2 | m2 | 0.02 | 0.30 | 0.12 | 0.06 | 0.48 |
| 30g per m2 | m2 | 0.02 | 0.30 | 0.13 | 0.06 | 0.49 |
| 32g per m2 | m2 | 0.02 | 0.30 | 0.14 | 0.07 | 0.51 |
| 34g per m2 | m2 | 0.02 | 0.30 | 0.14 | 0.07 | 0.51 |
| 36g per m2 | m2 | 0.02 | 0.30 | 0.15 | 0.07 | 0.52 |
| 38g per m2 | m2 | 0.02 | 0.30 | 0.16 | 0.07 | 0.53 |
| 40g per m2 | m2 | 0.03 | 0.45 | 0.17 | 0.09 | 0.71 |
| 42g per m2 | m2 | 0.03 | 0.45 | 0.18 | 0.09 | 0.72 |
| 44g per m2 | m2 | 0.03 | 0.45 | 0.18 | 0.09 | 0.72 |
| 46g per m2 | m2 | 0.03 | 0.45 | 0.19 | 0.10 | 0.74 |
| 48g per m2 | m2 | 0.03 | 0.45 | 0.20 | 0.10 | 0.75 |
| 50g per m2 | m2 | 0.03 | 0.45 | 0.21 | 0.10 | 0.76 |

| | Unit | Labour | Hours £ | Mat'ls £ | O & P £ | Total £ |
|---|---|---|---|---|---|---|
| **Grass seeding by hand (cont'd)** | | | | | | |
| PC £110.00 per 25kg | | | | | | |
| 10g per m2 | m2 | 0.01 | 0.15 | 0.05 | 0.03 | 0.23 |
| 12g per m2 | m2 | 0.01 | 0.15 | 0.06 | 0.03 | 0.24 |
| 14g per m2 | m2 | 0.01 | 0.15 | 0.07 | 0.03 | 0.25 |
| 16g per m2 | m2 | 0.01 | 0.15 | 0.07 | 0.03 | 0.25 |
| 18g per m2 | m2 | 0.01 | 0.15 | 0.08 | 0.03 | 0.26 |
| 20g per m2 | m2 | 0.01 | 0.15 | 0.09 | 0.04 | 0.28 |
| 22g per m2 | m2 | 0.01 | 0.15 | 0.10 | 0.04 | 0.29 |
| 24g per m2 | m2 | 0.01 | 0.15 | 0.11 | 0.04 | 0.30 |
| 26g per m2 | m2 | 0.02 | 0.30 | 0.11 | 0.06 | 0.47 |
| 28g per m2 | m2 | 0.02 | 0.30 | 0.12 | 0.06 | 0.48 |
| 30g per m2 | m2 | 0.02 | 0.30 | 0.13 | 0.06 | 0.49 |
| 32g per m2 | m2 | 0.02 | 0.30 | 0.14 | 0.07 | 0.51 |
| 34g per m2 | m2 | 0.02 | 0.30 | 0.14 | 0.07 | 0.51 |
| 36g per m2 | m2 | 0.02 | 0.30 | 0.15 | 0.07 | 0.52 |
| 38g per m2 | m2 | 0.02 | 0.30 | 0.16 | 0.07 | 0.53 |
| 40g per m2 | m2 | 0.03 | 0.45 | 0.17 | 0.09 | 0.71 |
| 42g per m2 | m2 | 0.03 | 0.45 | 0.18 | 0.09 | 0.72 |
| 44g per m2 | m2 | 0.03 | 0.45 | 0.18 | 0.09 | 0.72 |
| 46g per m2 | m2 | 0.03 | 0.45 | 0.19 | 0.10 | 0.74 |
| 48g per m2 | m2 | 0.03 | 0.45 | 0.20 | 0.10 | 0.75 |
| 50g per m2 | m2 | 0.03 | 0.45 | 0.21 | 0.10 | 0.76 |
| PC £120.00 per 25kg | | | | | | |
| 10g per m2 | m2 | 0.01 | 0.15 | 0.06 | 0.03 | 0.24 |
| 12g per m2 | m2 | 0.01 | 0.15 | 0.07 | 0.03 | 0.25 |
| 14g per m2 | m2 | 0.01 | 0.15 | 0.08 | 0.03 | 0.26 |
| 16g per m2 | m2 | 0.01 | 0.15 | 0.09 | 0.04 | 0.28 |
| 18g per m2 | m2 | 0.01 | 0.15 | 0.10 | 0.04 | 0.29 |
| 20g per m2 | m2 | 0.01 | 0.15 | 0.11 | 0.04 | 0.30 |
| 22g per m2 | m2 | 0.01 | 0.15 | 0.12 | 0.04 | 0.31 |
| 24g per m2 | m2 | 0.01 | 0.15 | 0.13 | 0.04 | 0.32 |
| 26g per m2 | m2 | 0.02 | 0.30 | 0.14 | 0.07 | 0.51 |

| | Unit | Labour | Hours £ | Mat'ls £ | O & P £ | Total £ |
|---|---|---|---|---|---|---|
| 28g per m2 | m2 | 0.02 | 0.30 | 0.15 | 0.07 | 0.52 |
| 30g per m2 | m2 | 0.02 | 0.30 | 0.16 | 0.07 | 0.53 |
| 32g per m2 | m2 | 0.02 | 0.30 | 0.17 | 0.07 | 0.54 |
| 34g per m2 | m2 | 0.02 | 0.30 | 0.18 | 0.07 | 0.55 |
| 36g per m2 | m2 | 0.02 | 0.30 | 0.19 | 0.07 | 0.56 |
| 38g per m2 | m2 | 0.02 | 0.30 | 0.20 | 0.08 | 0.58 |
| 40g per m2 | m2 | 0.03 | 0.45 | 0.21 | 0.10 | 0.76 |
| 42g per m2 | m2 | 0.03 | 0.45 | 0.22 | 0.10 | 0.77 |
| 44g per m2 | m2 | 0.03 | 0.45 | 0.23 | 0.10 | 0.78 |
| 46g per m2 | m2 | 0.03 | 0.45 | 0.24 | 0.10 | 0.79 |
| 48g per m2 | m2 | 0.03 | 0.45 | 0.25 | 0.11 | 0.81 |
| 50g per m2 | m2 | 0.03 | 0.45 | 0.26 | 0.11 | 0.82 |
| PC £130.00 per 25kg | | | | | | |
| 10g per m2 | m2 | 0.01 | 0.15 | 0.06 | 0.03 | 0.24 |
| 12g per m2 | m2 | 0.01 | 0.15 | 0.07 | 0.03 | 0.25 |
| 14g per m2 | m2 | 0.01 | 0.15 | 0.08 | 0.03 | 0.26 |
| 16g per m2 | m2 | 0.01 | 0.15 | 0.09 | 0.04 | 0.28 |
| 18g per m2 | m2 | 0.01 | 0.15 | 0.10 | 0.04 | 0.29 |
| 20g per m2 | m2 | 0.01 | 0.15 | 0.11 | 0.04 | 0.30 |
| 22g per m2 | m2 | 0.01 | 0.15 | 0.12 | 0.04 | 0.31 |
| 24g per m2 | m2 | 0.01 | 0.15 | 0.13 | 0.04 | 0.32 |
| 26g per m2 | m2 | 0.02 | 0.30 | 0.14 | 0.07 | 0.51 |
| 28g per m2 | m2 | 0.02 | 0.30 | 0.15 | 0.07 | 0.52 |
| 30g per m2 | m2 | 0.02 | 0.30 | 0.16 | 0.07 | 0.53 |
| 32g per m2 | m2 | 0.02 | 0.30 | 0.17 | 0.07 | 0.54 |
| 34g per m2 | m2 | 0.02 | 0.30 | 0.18 | 0.07 | 0.55 |
| 36g per m2 | m2 | 0.02 | 0.30 | 0.19 | 0.07 | 0.56 |
| 38g per m2 | m2 | 0.02 | 0.30 | 0.20 | 0.08 | 0.58 |
| 40g per m2 | m2 | 0.03 | 0.45 | 0.21 | 0.10 | 0.76 |
| 42g per m2 | m2 | 0.03 | 0.45 | 0.22 | 0.10 | 0.77 |
| 44g per m2 | m2 | 0.03 | 0.45 | 0.23 | 0.10 | 0.78 |
| 46g per m2 | m2 | 0.03 | 0.45 | 0.24 | 0.10 | 0.79 |
| 48g per m2 | m2 | 0.03 | 0.45 | 0.25 | 0.11 | 0.81 |
| 50g per m2 | m2 | 0.03 | 0.45 | 0.26 | 0.11 | 0.82 |

| | Unit | Labour | Hours £ | Mat'ls £ | O & P £ | Total £ |
|---|---|---|---|---|---|---|
| **Grass seeding by hand (cont'd)** | | | | | | |
| PC £140.00 per 25kg | | | | | | |
| 10g per m2 | m2 | 0.01 | 0.15 | 0.06 | 0.03 | 0.24 |
| 12g per m2 | m2 | 0.01 | 0.15 | 0.07 | 0.03 | 0.25 |
| 14g per m2 | m2 | 0.01 | 0.15 | 0.08 | 0.03 | 0.26 |
| 16g per m2 | m2 | 0.01 | 0.15 | 0.09 | 0.04 | 0.28 |
| 18g per m2 | m2 | 0.01 | 0.15 | 0.10 | 0.04 | 0.29 |
| 20g per m2 | m2 | 0.01 | 0.15 | 0.11 | 0.04 | 0.30 |
| 22g per m2 | m2 | 0.01 | 0.15 | 0.12 | 0.04 | 0.31 |
| 24g per m2 | m2 | 0.01 | 0.15 | 0.13 | 0.04 | 0.32 |
| 26g per m2 | m2 | 0.02 | 0.30 | 0.14 | 0.07 | 0.51 |
| 28g per m2 | m2 | 0.02 | 0.30 | 0.15 | 0.07 | 0.52 |
| 30g per m2 | m2 | 0.02 | 0.30 | 0.16 | 0.07 | 0.53 |
| 32g per m2 | m2 | 0.02 | 0.30 | 0.17 | 0.07 | 0.54 |
| 34g per m2 | m2 | 0.02 | 0.30 | 0.18 | 0.07 | 0.55 |
| 36g per m2 | m2 | 0.02 | 0.30 | 0.19 | 0.07 | 0.56 |
| 38g per m2 | m2 | 0.02 | 0.30 | 0.20 | 0.08 | 0.58 |
| 40g per m2 | m2 | 0.03 | 0.45 | 0.21 | 0.10 | 0.76 |
| 42g per m2 | m2 | 0.03 | 0.45 | 0.22 | 0.10 | 0.77 |
| 44g per m2 | m2 | 0.03 | 0.45 | 0.23 | 0.10 | 0.78 |
| 46g per m2 | m2 | 0.03 | 0.45 | 0.24 | 0.10 | 0.79 |
| 48g per m2 | m2 | 0.03 | 0.45 | 0.25 | 0.11 | 0.81 |
| 50g per m2 | m2 | 0.03 | 0.45 | 0.26 | 0.11 | 0.82 |
| PC £150.00 per 25kg | | | | | | |
| 10g per m2 | m2 | 0.01 | 0.15 | 0.07 | 0.03 | 0.25 |
| 12g per m2 | m2 | 0.01 | 0.15 | 0.08 | 0.03 | 0.26 |
| 14g per m2 | m2 | 0.01 | 0.15 | 0.09 | 0.04 | 0.28 |
| 16g per m2 | m2 | 0.01 | 0.15 | 0.11 | 0.04 | 0.30 |
| 18g per m2 | m2 | 0.01 | 0.15 | 0.12 | 0.04 | 0.31 |
| 20g per m2 | m2 | 0.01 | 0.15 | 0.13 | 0.04 | 0.32 |
| 22g per m2 | m2 | 0.01 | 0.15 | 0.14 | 0.04 | 0.33 |
| 24g per m2 | m2 | 0.01 | 0.15 | 0.15 | 0.05 | 0.35 |
| 26g per m2 | m2 | 0.02 | 0.30 | 0.16 | 0.07 | 0.53 |

| | Unit | Labour | Hours £ | Mat'ls £ | O & P £ | Total £ |
|---|---|---|---|---|---|---|
| 28g per m2 | m2 | 0.02 | 0.30 | 0.17 | 0.07 | 0.54 |
| 30g per m2 | m2 | 0.02 | 0.30 | 0.18 | 0.07 | 0.55 |
| 32g per m2 | m2 | 0.02 | 0.30 | 0.19 | 0.07 | 0.56 |
| 34g per m2 | m2 | 0.02 | 0.30 | 0.21 | 0.08 | 0.59 |
| 36g per m2 | m2 | 0.02 | 0.30 | 0.22 | 0.08 | 0.60 |
| 38g per m2 | m2 | 0.02 | 0.30 | 0.23 | 0.08 | 0.61 |
| 40g per m2 | m2 | 0.03 | 0.45 | 0.25 | 0.11 | 0.81 |
| 42g per m2 | m2 | 0.03 | 0.45 | 0.26 | 0.11 | 0.82 |
| 44g per m2 | m2 | 0.03 | 0.45 | 0.27 | 0.11 | 0.83 |
| 46g per m2 | m2 | 0.03 | 0.45 | 0.28 | 0.11 | 0.84 |
| 48g per m2 | m2 | 0.03 | 0.45 | 0.30 | 0.11 | 0.86 |
| 50g per m2 | m2 | 0.03 | 0.45 | 0.31 | 0.11 | 0.87 |
| PC £160.00 per 25kg | | | | | | |
| 10g per m2 | m2 | 0.01 | 0.15 | 0.07 | 0.03 | 0.25 |
| 12g per m2 | m2 | 0.01 | 0.15 | 0.08 | 0.03 | 0.26 |
| 14g per m2 | m2 | 0.01 | 0.15 | 0.09 | 0.04 | 0.28 |
| 16g per m2 | m2 | 0.01 | 0.15 | 0.11 | 0.04 | 0.30 |
| 18g per m2 | m2 | 0.01 | 0.15 | 0.12 | 0.04 | 0.31 |
| 20g per m2 | m2 | 0.01 | 0.15 | 0.13 | 0.04 | 0.32 |
| 22g per m2 | m2 | 0.01 | 0.15 | 0.14 | 0.04 | 0.33 |
| 24g per m2 | m2 | 0.01 | 0.15 | 0.15 | 0.05 | 0.35 |
| 26g per m2 | m2 | 0.02 | 0.30 | 0.16 | 0.07 | 0.53 |
| 28g per m2 | m2 | 0.02 | 0.30 | 0.17 | 0.07 | 0.54 |
| 30g per m2 | m2 | 0.02 | 0.30 | 0.18 | 0.07 | 0.55 |
| 32g per m2 | m2 | 0.02 | 0.30 | 0.19 | 0.07 | 0.56 |
| 34g per m2 | m2 | 0.02 | 0.30 | 0.21 | 0.08 | 0.59 |
| 36g per m2 | m2 | 0.02 | 0.30 | 0.22 | 0.08 | 0.60 |
| 38g per m2 | m2 | 0.02 | 0.30 | 0.23 | 0.08 | 0.61 |
| 40g per m2 | m2 | 0.03 | 0.45 | 0.25 | 0.11 | 0.81 |
| 42g per m2 | m2 | 0.03 | 0.45 | 0.26 | 0.11 | 0.82 |
| 44g per m2 | m2 | 0.03 | 0.45 | 0.27 | 0.11 | 0.83 |
| 46g per m2 | m2 | 0.03 | 0.45 | 0.28 | 0.11 | 0.84 |
| 48g per m2 | m2 | 0.03 | 0.45 | 0.30 | 0.11 | 0.86 |
| 50g per m2 | m2 | 0.03 | 0.45 | 0.31 | 0.11 | 0.87 |

| | Unit | Labour | Hours £ | Mat'ls £ | O & P £ | Total £ |
|---|---|---|---|---|---|---|
| **Lawn treatment by hand by hand** | | | | | | |
| Apply lawn treatment | | | | | | |
| bonemeal | m2 | 0.02 | 0.30 | 0.03 | 0.05 | 0.38 |
| fish, blood and stone | m2 | 0.02 | 0.30 | 0.03 | 0.05 | 0.38 |
| autumn/winter feed | m2 | 0.02 | 0.30 | 0.04 | 0.05 | 0.39 |
| spring/summer feed | m2 | 0.02 | 0.30 | 0.04 | 0.05 | 0.39 |
| weed and moss killer | m2 | 0.02 | 0.30 | 0.05 | 0.05 | 0.40 |
| **Turfing by hand** | | | | | | |
| Lay imported turf on prepared bed | | | | | | |
| meadow turf | m2 | 0.08 | 1.20 | 1.22 | 0.36 | 2.78 |
| sports outfields | m2 | 0.09 | 1.35 | 1.82 | 0.48 | 3.65 |
| domestic lawns | m2 | 0.10 | 1.50 | 2.09 | 0.54 | 4.13 |

| | Unit | Plant £ | Mat'ls £ | O & P £ | Total £ |
|---|---|---|---|---|---|
| **Grass seeding by machine** | | | | | |
| Where applicable the plant column includes the cost of the operator. | | | | | |
| Sow grass seed on prepared ground | | | | | |
| PC £60.00 per 25kg | | | | | |
| 10g per m2 | m2 | 0.01 | 0.03 | 0.01 | 0.05 |
| 12g per m2 | m2 | 0.01 | 0.03 | 0.01 | 0.05 |
| 14g per m2 | m2 | 0.01 | 0.04 | 0.01 | 0.06 |
| 16g per m2 | m2 | 0.01 | 0.04 | 0.01 | 0.06 |
| 18g per m2 | m2 | 0.01 | 0.04 | 0.01 | 0.06 |
| 20g per m2 | m2 | 0.01 | 0.05 | 0.01 | 0.07 |
| 22g per m2 | m2 | 0.01 | 0.05 | 0.01 | 0.07 |
| 24g per m2 | m2 | 0.01 | 0.05 | 0.01 | 0.07 |
| 26g per m2 | m2 | 0.01 | 0.06 | 0.01 | 0.08 |
| 28g per m2 | m2 | 0.01 | 0.06 | 0.01 | 0.08 |
| 30g per m2 | m2 | 0.01 | 0.07 | 0.01 | 0.09 |
| 32g per m2 | m2 | 0.01 | 0.07 | 0.01 | 0.09 |
| 34g per m2 | m2 | 0.01 | 0.07 | 0.01 | 0.09 |
| 36g per m2 | m2 | 0.01 | 0.08 | 0.01 | 0.10 |
| 38g per m2 | m2 | 0.01 | 0.08 | 0.01 | 0.10 |
| 40g per m2 | m2 | 0.01 | 0.09 | 0.02 | 0.12 |
| 42g per m2 | m2 | 0.01 | 0.09 | 0.02 | 0.12 |
| 44g per m2 | m2 | 0.01 | 0.09 | 0.02 | 0.12 |
| 46g per m2 | m2 | 0.01 | 0.10 | 0.02 | 0.13 |
| 48g per m2 | m2 | 0.01 | 0.10 | 0.02 | 0.13 |
| 50g per m2 | m2 | 0.01 | 0.10 | 0.02 | 0.13 |
| PC £70.00 per 25kg | | | | | |
| 10g per m2 | m2 | 0.01 | 0.04 | 0.01 | 0.06 |
| 12g per m2 | m2 | 0.01 | 0.04 | 0.01 | 0.06 |
| 14g per m2 | m2 | 0.01 | 0.05 | 0.01 | 0.07 |

| | Unit | Plant £ | Mat'ls £ | O & P £ | Total £ |
|---|---|---|---|---|---|
| **Grass seeding by machine (cont'd)** | | | | | |
| 16g per m2 | m2 | 0.01 | 0.05 | 0.01 | 0.07 |
| 18g per m2 | m2 | 0.01 | 0.06 | 0.01 | 0.08 |
| 20g per m2 | m2 | 0.01 | 0.07 | 0.01 | 0.09 |
| 22g per m2 | m2 | 0.01 | 0.07 | 0.01 | 0.09 |
| 24g per m2 | m2 | 0.01 | 0.08 | 0.01 | 0.10 |
| 26g per m2 | m2 | 0.01 | 0.09 | 0.02 | 0.12 |
| 28g per m2 | m2 | 0.01 | 0.09 | 0.02 | 0.12 |
| 30g per m2 | m2 | 0.01 | 0.10 | 0.02 | 0.13 |
| 32g per m2 | m2 | 0.01 | 0.10 | 0.02 | 0.13 |
| 34g per m2 | m2 | 0.01 | 0.11 | 0.02 | 0.14 |
| 36g per m2 | m2 | 0.01 | 0.12 | 0.02 | 0.15 |
| 38g per m2 | m2 | 0.01 | 0.12 | 0.02 | 0.15 |
| 40g per m2 | m2 | 0.01 | 0.13 | 0.02 | 0.16 |
| 42g per m2 | m2 | 0.01 | 0.13 | 0.02 | 0.16 |
| 44g per m2 | m2 | 0.01 | 0.14 | 0.02 | 0.17 |
| 46g per m2 | m2 | 0.01 | 0.14 | 0.02 | 0.17 |
| 48g per m2 | m2 | 0.01 | 0.15 | 0.02 | 0.18 |
| 50g per m2 | m2 | 0.01 | 0.16 | 0.03 | 0.20 |
| PC £80.00 per 25kg | | | | | |
| 10g per m2 | m2 | 0.01 | 0.04 | 0.01 | 0.06 |
| 12g per m2 | m2 | 0.01 | 0.04 | 0.01 | 0.06 |
| 14g per m2 | m2 | 0.01 | 0.05 | 0.01 | 0.07 |
| 16g per m2 | m2 | 0.01 | 0.05 | 0.01 | 0.07 |
| 18g per m2 | m2 | 0.01 | 0.06 | 0.01 | 0.08 |
| 20g per m2 | m2 | 0.01 | 0.07 | 0.01 | 0.09 |
| 22g per m2 | m2 | 0.01 | 0.07 | 0.01 | 0.09 |
| 24g per m2 | m2 | 0.01 | 0.08 | 0.01 | 0.10 |
| 26g per m2 | m2 | 0.01 | 0.09 | 0.02 | 0.12 |
| 28g per m2 | m2 | 0.01 | 0.09 | 0.02 | 0.12 |
| 30g per m2 | m2 | 0.01 | 0.10 | 0.02 | 0.13 |
| 32g per m2 | m2 | 0.01 | 0.10 | 0.02 | 0.13 |
| 34g per m2 | m2 | 0.01 | 0.11 | 0.02 | 0.14 |

| | Unit | Plant £ | Mat'ls £ | O & P £ | Total £ |
|---|---|---|---|---|---|
| 36g per m2 | m2 | 0.01 | 0.12 | 0.02 | 0.15 |
| 38g per m2 | m2 | 0.01 | 0.12 | 0.02 | 0.15 |
| 40g per m2 | m2 | 0.01 | 0.13 | 0.02 | 0.16 |
| 42g per m2 | m2 | 0.01 | 0.13 | 0.02 | 0.16 |
| 44g per m2 | m2 | 0.01 | 0.14 | 0.02 | 0.17 |
| 46g per m2 | m2 | 0.01 | 0.14 | 0.02 | 0.17 |
| 48g per m2 | m2 | 0.01 | 0.15 | 0.02 | 0.18 |
| 50g per m2 | m2 | 0.01 | 0.16 | 0.03 | 0.20 |
| PC £90.00 per 25kg | | | | | |
| 10g per m2 | m2 | 0.01 | 0.04 | 0.01 | 0.06 |
| 12g per m2 | m2 | 0.01 | 0.04 | 0.01 | 0.06 |
| 14g per m2 | m2 | 0.01 | 0.05 | 0.01 | 0.07 |
| 16g per m2 | m2 | 0.01 | 0.05 | 0.01 | 0.07 |
| 18g per m2 | m2 | 0.01 | 0.06 | 0.01 | 0.08 |
| 20g per m2 | m2 | 0.01 | 0.07 | 0.01 | 0.09 |
| 22g per m2 | m2 | 0.01 | 0.07 | 0.01 | 0.09 |
| 24g per m2 | m2 | 0.01 | 0.08 | 0.01 | 0.10 |
| 26g per m2 | m2 | 0.01 | 0.09 | 0.02 | 0.12 |
| 28g per m2 | m2 | 0.01 | 0.09 | 0.02 | 0.12 |
| 30g per m2 | m2 | 0.01 | 0.10 | 0.02 | 0.13 |
| 32g per m2 | m2 | 0.01 | 0.10 | 0.02 | 0.13 |
| 34g per m2 | m2 | 0.01 | 0.11 | 0.02 | 0.14 |
| 36g per m2 | m2 | 0.01 | 0.12 | 0.02 | 0.15 |
| 38g per m2 | m2 | 0.01 | 0.12 | 0.02 | 0.15 |
| 40g per m2 | m2 | 0.01 | 0.13 | 0.02 | 0.16 |
| 42g per m2 | m2 | 0.01 | 0.13 | 0.02 | 0.16 |
| 44g per m2 | m2 | 0.01 | 0.14 | 0.02 | 0.17 |
| 46g per m2 | m2 | 0.01 | 0.14 | 0.02 | 0.17 |
| 48g per m2 | m2 | 0.01 | 0.15 | 0.02 | 0.18 |
| 50g per m2 | m2 | 0.01 | 0.16 | 0.03 | 0.20 |

| | Unit | Plant £ | Mat'ls £ | O & P £ | Total £ |
|---|---|---|---|---|---|
| **Grass seeding by machine (cont'd)** | | | | | |
| PC £100.00 per 25kg | | | | | |
| 10g per m2 | m2 | 0.01 | 0.05 | 0.01 | 0.07 |
| 12g per m2 | m2 | 0.01 | 0.06 | 0.01 | 0.08 |
| 14g per m2 | m2 | 0.01 | 0.07 | 0.01 | 0.09 |
| 16g per m2 | m2 | 0.01 | 0.07 | 0.01 | 0.09 |
| 18g per m2 | m2 | 0.01 | 0.08 | 0.01 | 0.10 |
| 20g per m2 | m2 | 0.01 | 0.09 | 0.02 | 0.12 |
| 22g per m2 | m2 | 0.01 | 0.10 | 0.02 | 0.13 |
| 24g per m2 | m2 | 0.01 | 0.11 | 0.02 | 0.14 |
| 26g per m2 | m2 | 0.01 | 0.11 | 0.02 | 0.14 |
| 28g per m2 | m2 | 0.01 | 0.12 | 0.02 | 0.15 |
| 30g per m2 | m2 | 0.01 | 0.13 | 0.02 | 0.16 |
| 32g per m2 | m2 | 0.01 | 0.14 | 0.02 | 0.17 |
| 34g per m2 | m2 | 0.01 | 0.14 | 0.02 | 0.17 |
| 36g per m2 | m2 | 0.01 | 0.15 | 0.02 | 0.18 |
| 38g per m2 | m2 | 0.01 | 0.16 | 0.03 | 0.20 |
| 40g per m2 | m2 | 0.01 | 0.17 | 0.03 | 0.21 |
| 42g per m2 | m2 | 0.01 | 0.18 | 0.03 | 0.22 |
| 44g per m2 | m2 | 0.01 | 0.18 | 0.03 | 0.22 |
| 46g per m2 | m2 | 0.01 | 0.19 | 0.03 | 0.23 |
| 48g per m2 | m2 | 0.01 | 0.20 | 0.03 | 0.24 |
| 50g per m2 | m2 | 0.01 | 0.21 | 0.03 | 0.25 |
| PC £110.00 per 25kg | | | | | |
| 10g per m2 | m2 | 0.01 | 0.05 | 0.01 | 0.07 |
| 12g per m2 | m2 | 0.01 | 0.06 | 0.01 | 0.08 |
| 14g per m2 | m2 | 0.01 | 0.07 | 0.01 | 0.09 |
| 16g per m2 | m2 | 0.01 | 0.07 | 0.01 | 0.09 |
| 18g per m2 | m2 | 0.01 | 0.08 | 0.01 | 0.10 |
| 20g per m2 | m2 | 0.01 | 0.09 | 0.02 | 0.12 |
| 22g per m2 | m2 | 0.01 | 0.10 | 0.02 | 0.13 |
| 24g per m2 | m2 | 0.01 | 0.11 | 0.02 | 0.14 |
| 26g per m2 | m2 | 0.01 | 0.11 | 0.02 | 0.14 |

| | Unit | Plant £ | Mat'ls £ | O & P £ | Total £ |
|---|---|---|---|---|---|
| 28g per m2 | m2 | 0.01 | 0.12 | 0.02 | 0.15 |
| 30g per m2 | m2 | 0.01 | 0.13 | 0.02 | 0.16 |
| 32g per m2 | m2 | 0.01 | 0.14 | 0.02 | 0.17 |
| 34g per m2 | m2 | 0.01 | 0.14 | 0.02 | 0.17 |
| 36g per m2 | m2 | 0.01 | 0.15 | 0.02 | 0.18 |
| 38g per m2 | m2 | 0.01 | 0.16 | 0.03 | 0.20 |
| 40g per m2 | m2 | 0.01 | 0.17 | 0.03 | 0.21 |
| 42g per m2 | m2 | 0.01 | 0.18 | 0.03 | 0.22 |
| 44g per m2 | m2 | 0.01 | 0.18 | 0.03 | 0.22 |
| 46g per m2 | m2 | 0.01 | 0.19 | 0.03 | 0.23 |
| 48g per m2 | m2 | 0.01 | 0.20 | 0.03 | 0.24 |
| 50g per m2 | m2 | 0.01 | 0.21 | 0.03 | 0.25 |
| PC £120.00 per 25kg | | | | | |
| 10g per m2 | m2 | 0.01 | 0.06 | 0.01 | 0.08 |
| 12g per m2 | m2 | 0.01 | 0.07 | 0.01 | 0.09 |
| 14g per m2 | m2 | 0.01 | 0.08 | 0.01 | 0.10 |
| 16g per m2 | m2 | 0.01 | 0.09 | 0.02 | 0.12 |
| 18g per m2 | m2 | 0.01 | 0.10 | 0.02 | 0.13 |
| 20g per m2 | m2 | 0.01 | 0.11 | 0.02 | 0.14 |
| 22g per m2 | m2 | 0.01 | 0.12 | 0.02 | 0.15 |
| 24g per m2 | m2 | 0.01 | 0.13 | 0.02 | 0.16 |
| 26g per m2 | m2 | 0.01 | 0.14 | 0.02 | 0.17 |
| 28g per m2 | m2 | 0.01 | 0.15 | 0.02 | 0.18 |
| 30g per m2 | m2 | 0.01 | 0.16 | 0.03 | 0.20 |
| 32g per m2 | m2 | 0.01 | 0.17 | 0.03 | 0.21 |
| 34g per m2 | m2 | 0.01 | 0.18 | 0.03 | 0.22 |
| 36g per m2 | m2 | 0.01 | 0.19 | 0.03 | 0.23 |
| 38g per m2 | m2 | 0.01 | 0.20 | 0.03 | 0.24 |
| 40g per m2 | m2 | 0.01 | 0.21 | 0.03 | 0.25 |
| 42g per m2 | m2 | 0.01 | 0.22 | 0.03 | 0.26 |
| 44g per m2 | m2 | 0.01 | 0.23 | 0.04 | 0.28 |
| 46g per m2 | m2 | 0.01 | 0.24 | 0.04 | 0.29 |
| 48g per m2 | m2 | 0.01 | 0.25 | 0.04 | 0.30 |
| 50g per m2 | m2 | 0.01 | 0.26 | 0.04 | 0.31 |

| | Unit | Plant £ | Mat'ls £ | O & P £ | Total £ |
|---|---|---|---|---|---|
| **Grass seeding by machine (cont'd)** | | | | | |
| PC £130.00 per 25kg | | | | | |
| 10g per m2 | m2 | 0.01 | 0.06 | 0.01 | 0.08 |
| 12g per m2 | m2 | 0.01 | 0.07 | 0.01 | 0.09 |
| 14g per m2 | m2 | 0.01 | 0.08 | 0.01 | 0.10 |
| 16g per m2 | m2 | 0.01 | 0.09 | 0.02 | 0.12 |
| 18g per m2 | m2 | 0.01 | 0.10 | 0.02 | 0.13 |
| 20g per m2 | m2 | 0.01 | 0.11 | 0.02 | 0.14 |
| 22g per m2 | m2 | 0.01 | 0.12 | 0.02 | 0.15 |
| 24g per m2 | m2 | 0.01 | 0.13 | 0.02 | 0.16 |
| 26g per m2 | m2 | 0.01 | 0.14 | 0.02 | 0.17 |
| 28g per m2 | m2 | 0.01 | 0.15 | 0.02 | 0.18 |
| 30g per m2 | m2 | 0.01 | 0.16 | 0.03 | 0.20 |
| 32g per m2 | m2 | 0.01 | 0.17 | 0.03 | 0.21 |
| 34g per m2 | m2 | 0.01 | 0.18 | 0.03 | 0.22 |
| 36g per m2 | m2 | 0.01 | 0.19 | 0.03 | 0.23 |
| 38g per m2 | m2 | 0.01 | 0.20 | 0.03 | 0.24 |
| 40g per m2 | m2 | 0.01 | 0.21 | 0.03 | 0.25 |
| 42g per m2 | m2 | 0.01 | 0.22 | 0.03 | 0.26 |
| 44g per m2 | m2 | 0.01 | 0.23 | 0.04 | 0.28 |
| 46g per m2 | m2 | 0.01 | 0.24 | 0.04 | 0.29 |
| 48g per m2 | m2 | 0.01 | 0.25 | 0.04 | 0.30 |
| 50g per m2 | m2 | 0.01 | 0.26 | 0.04 | 0.31 |
| PC £140.00 per 25kg | | | | | |
| 10g per m2 | m2 | 0.01 | 0.06 | 0.01 | 0.08 |
| 12g per m2 | m2 | 0.01 | 0.07 | 0.01 | 0.09 |
| 14g per m2 | m2 | 0.01 | 0.08 | 0.01 | 0.10 |
| 16g per m2 | m2 | 0.01 | 0.09 | 0.02 | 0.12 |
| 18g per m2 | m2 | 0.01 | 0.10 | 0.02 | 0.13 |
| 20g per m2 | m2 | 0.01 | 0.11 | 0.02 | 0.14 |
| 22g per m2 | m2 | 0.01 | 0.12 | 0.02 | 0.15 |
| 24g per m2 | m2 | 0.01 | 0.13 | 0.02 | 0.16 |
| 26g per m2 | m2 | 0.01 | 0.14 | 0.02 | 0.17 |

| | Unit | Plant £ | Mat'ls £ | O & P £ | Total £ |
|---|---|---|---|---|---|
| 28g per m2 | m2 | 0.01 | 0.15 | 0.02 | 0.18 |
| 30g per m2 | m2 | 0.01 | 0.16 | 0.03 | 0.20 |
| 32g per m2 | m2 | 0.01 | 0.17 | 0.03 | 0.21 |
| 34g per m2 | m2 | 0.01 | 0.18 | 0.03 | 0.22 |
| 36g per m2 | m2 | 0.01 | 0.19 | 0.03 | 0.23 |
| 38g per m2 | m2 | 0.01 | 0.20 | 0.03 | 0.24 |
| 40g per m2 | m2 | 0.01 | 0.21 | 0.03 | 0.25 |
| 42g per m2 | m2 | 0.01 | 0.22 | 0.03 | 0.26 |
| 44g per m2 | m2 | 0.01 | 0.23 | 0.04 | 0.28 |
| 46g per m2 | m2 | 0.01 | 0.24 | 0.04 | 0.29 |
| 48g per m2 | m2 | 0.01 | 0.25 | 0.04 | 0.30 |
| 50g per m2 | m2 | 0.01 | 0.26 | 0.04 | 0.31 |
| PC £150.00 per 25kg | | | | | |
| 10g per m2 | m2 | 0.01 | 0.07 | 0.01 | 0.09 |
| 12g per m2 | m2 | 0.01 | 0.08 | 0.01 | 0.10 |
| 14g per m2 | m2 | 0.01 | 0.09 | 0.02 | 0.12 |
| 16g per m2 | m2 | 0.01 | 0.11 | 0.02 | 0.14 |
| 18g per m2 | m2 | 0.01 | 0.12 | 0.02 | 0.15 |
| 20g per m2 | m2 | 0.01 | 0.13 | 0.02 | 0.16 |
| 22g per m2 | m2 | 0.01 | 0.14 | 0.02 | 0.17 |
| 24g per m2 | m2 | 0.01 | 0.15 | 0.02 | 0.18 |
| 26g per m2 | m2 | 0.01 | 0.16 | 0.03 | 0.20 |
| 28g per m2 | m2 | 0.01 | 0.17 | 0.03 | 0.21 |
| 30g per m2 | m2 | 0.01 | 0.18 | 0.03 | 0.22 |
| 32g per m2 | m2 | 0.01 | 0.19 | 0.03 | 0.23 |
| 34g per m2 | m2 | 0.01 | 0.21 | 0.03 | 0.25 |
| 36g per m2 | m2 | 0.01 | 0.22 | 0.03 | 0.26 |
| 38g per m2 | m2 | 0.01 | 0.23 | 0.04 | 0.28 |
| 40g per m2 | m2 | 0.01 | 0.25 | 0.04 | 0.30 |
| 42g per m2 | m2 | 0.01 | 0.26 | 0.04 | 0.31 |
| 44g per m2 | m2 | 0.01 | 0.27 | 0.04 | 0.32 |
| 46g per m2 | m2 | 0.01 | 0.28 | 0.04 | 0.33 |
| 48g per m2 | m2 | 0.01 | 0.30 | 0.05 | 0.36 |
| 50g per m2 | m2 | 0.01 | 0.31 | 0.05 | 0.37 |

| | Unit | Plant £ | Mat'ls £ | O & P £ | Total £ |
|---|---|---|---|---|---|
| **Grass seeding by machine (cont'd)** | | | | | |
| PC £160.00 per 25kg | | | | | |
| 10g per m2 | m2 | 0.01 | 0.07 | 0.01 | 0.09 |
| 12g per m2 | m2 | 0.01 | 0.08 | 0.01 | 0.10 |
| 14g per m2 | m2 | 0.01 | 0.09 | 0.02 | 0.12 |
| 16g per m2 | m2 | 0.01 | 0.11 | 0.02 | 0.14 |
| 18g per m2 | m2 | 0.01 | 0.12 | 0.02 | 0.15 |
| 20g per m2 | m2 | 0.01 | 0.13 | 0.02 | 0.16 |
| 22g per m2 | m2 | 0.01 | 0.14 | 0.02 | 0.17 |
| 24g per m2 | m2 | 0.01 | 0.15 | 0.02 | 0.18 |
| 26g per m2 | m2 | 0.01 | 0.16 | 0.03 | 0.20 |
| 28g per m2 | m2 | 0.01 | 0.17 | 0.03 | 0.21 |
| 30g per m2 | m2 | 0.01 | 0.18 | 0.03 | 0.22 |
| 32g per m2 | m2 | 0.01 | 0.19 | 0.03 | 0.23 |
| 34g per m2 | m2 | 0.01 | 0.21 | 0.03 | 0.25 |
| 36g per m2 | m2 | 0.01 | 0.22 | 0.03 | 0.26 |
| 38g per m2 | m2 | 0.01 | 0.23 | 0.04 | 0.28 |
| 40g per m2 | m2 | 0.01 | 0.25 | 0.04 | 0.30 |
| 42g per m2 | m2 | 0.01 | 0.26 | 0.04 | 0.31 |
| 44g per m2 | m2 | 0.01 | 0.27 | 0.04 | 0.32 |
| 46g per m2 | m2 | 0.01 | 0.28 | 0.04 | 0.33 |
| 48g per m2 | m2 | 0.01 | 0.30 | 0.05 | 0.36 |
| 50g per m2 | m2 | 0.01 | 0.31 | 0.05 | 0.37 |
| **Lawn treatment by machine by hand** | | | | | |
| Apply lawn treatment | | | | | |
| bonemeal | m2 | 0.01 | 0.03 | 0.01 | 0.05 |
| fish, blood and stone | m2 | 0.01 | 0.03 | 0.01 | 0.05 |
| autumn and winter feed | m2 | 0.01 | 0.04 | 0.01 | 0.06 |
| spring amd summer feed | m2 | 0.01 | 0.04 | 0.01 | 0.06 |
| weed and moss killer | m2 | 0.01 | 0.05 | 0.01 | 0.07 |

## BARE ROOT TREES

**The following descriptions refer to dimensions and sizes of trees as as set out below.**

Whips: 125 to 175cm high
Feathered: 175 to 250cm high
Light standard: 6 to 8cm girth
Standard: 8 to 10cm girth
Selected standard: 10 to 12cm girth
Heavy standard: 12 to 14cm girth
Extra heavy standard: 14 to 16cm girth

**Excavate tree pit, fork bottom, plant tree, backfill with excavated material including organic manure (30% of soil by volume), water and surround with peat**

| | Unit | Labour | Hours £ | Mat'ls £ | O & P £ | Total £ |
|---|---|---|---|---|---|---|
| Acer (Maple) | | | | | | |
| campestre | | | | | | |
| whip | nr | 0.30 | 4.50 | 4.06 | 1.28 | 9.84 |
| feathered | nr | 0.40 | 6.00 | 8.71 | 2.21 | 16.92 |
| light standard | nr | 1.00 | 15.00 | 17.42 | 4.86 | 37.28 |
| standard | nr | 1.20 | 18.00 | 19.92 | 5.69 | 43.61 |
| selected standard | nr | 1.40 | 21.00 | 34.20 | 8.28 | 63.48 |
| heavy standard | nr | 1.60 | 24.00 | 56.01 | 12.00 | 92.01 |
| extra heavy standard | nr | 2.00 | 30.00 | 68.42 | 14.76 | 113.18 |
| rubrum | | | | | | |
| whip | nr | 0.30 | 4.50 | 5.90 | 1.56 | 11.96 |
| feathered | nr | 0.40 | 6.00 | 12.45 | 2.77 | 21.22 |

| | Unit | Labour | Hours £ | Mat'ls £ | O & P £ | Total £ |
|---|---|---|---|---|---|---|
| **Tree planting (cont'd)** | | | | | | |
| light standard | nr | 1.00 | 15.00 | 24.25 | 5.89 | 45.14 |
| standard | nr | 1.20 | 18.00 | 34.21 | 7.83 | 60.04 |
| selected standard | nr | 1.40 | 21.00 | 56.02 | 11.55 | 88.57 |
| heavy standard | nr | 1.60 | 24.00 | 68.47 | 13.87 | 106.34 |
| extra heavy standard | nr | 2.00 | 30.00 | 80.91 | 16.64 | 127.55 |
| davidii | | | | | | |
| light standard | nr | 1.00 | 15.00 | 37.35 | 7.85 | 60.20 |
| standard | nr | 1.20 | 18.00 | 49.79 | 10.17 | 77.96 |
| selected standard | nr | 1.40 | 21.00 | 68.20 | 13.38 | 102.58 |
| heavy standard | nr | 1.60 | 24.00 | 99.59 | 18.54 | 142.13 |
| extra heavy standard | nr | 2.00 | 30.00 | 143.16 | 25.97 | 199.13 |
| platanoides | | | | | | |
| whip | nr | 0.30 | 4.50 | 2.16 | 1.00 | 7.66 |
| feathered | nr | 0.40 | 6.00 | 5.63 | 1.74 | 13.37 |
| light standard | nr | 1.00 | 15.00 | 11.53 | 3.98 | 30.51 |
| standard | nr | 1.20 | 18.00 | 14.02 | 4.80 | 36.82 |
| selected standard | nr | 1.40 | 21.00 | 20.24 | 6.19 | 47.43 |
| heavy standard | nr | 1.60 | 24.00 | 29.88 | 8.08 | 61.96 |
| extra heavy standard | nr | 2.00 | 30.00 | 68.47 | 14.77 | 113.24 |
| Aesculus (Chestnut) | | | | | | |
| hippocastanum | | | | | | |
| whip | nr | 0.30 | 4.50 | 4.06 | 1.28 | 9.84 |
| feathered | nr | 0.40 | 6.00 | 8.71 | 2.21 | 16.92 |
| light standard | nr | 1.00 | 15.00 | 17.43 | 4.86 | 37.29 |
| standard | nr | 1.20 | 18.00 | 19.92 | 5.69 | 43.61 |
| selected standard | nr | 1.40 | 21.00 | 34.21 | 8.28 | 63.49 |
| heavy standard | nr | 1.60 | 24.00 | 56.02 | 12.00 | 92.02 |
| extra heavy standard | nr | 2.00 | 30.00 | 68.20 | 14.73 | 112.93 |

| | Unit | Labour | Hours £ | Mat'ls £ | O & P £ | Total £ |
|---|---|---|---|---|---|---|
| h. baumanii | | | | | | |
| light standard | nr | 1.00 | 15.00 | 37.34 | 7.85 | 60.19 |
| standard | nr | 1.20 | 18.00 | 49.79 | 10.17 | 77.96 |
| selected standard | nr | 1.40 | 21.00 | 68.47 | 13.42 | 102.89 |
| heavy standard | nr | 1.60 | 24.00 | 99.59 | 18.54 | 142.13 |
| extra heavy standard | nr | 2.00 | 30.00 | 143.16 | 25.97 | 199.13 |
| Alnus (Alder) | | | | | | |
| glutinosa | | | | | | |
| whip | nr | 0.30 | 4.50 | 2.81 | 1.10 | 8.41 |
| feathered | nr | 0.40 | 6.00 | 6.82 | 1.92 | 14.74 |
| light standard | nr | 1.00 | 15.00 | 14.61 | 4.44 | 34.05 |
| standard | nr | 1.20 | 18.00 | 16.83 | 5.22 | 40.05 |
| selected standard | nr | 1.40 | 21.00 | 27.39 | 7.26 | 55.65 |
| heavy standard | nr | 1.60 | 24.00 | 49.79 | 11.07 | 84.86 |
| extra heavy standard | nr | 2.00 | 30.00 | 62.24 | 13.84 | 106.08 |
| g. laciniata | | | | | | |
| light standard | nr | 1.00 | 15.00 | 37.34 | 7.85 | 60.19 |
| standard | nr | 1.20 | 18.00 | 49.79 | 10.17 | 77.96 |
| selected standard | nr | 1.40 | 21.00 | 68.47 | 13.42 | 102.89 |
| heavy standard | nr | 1.60 | 24.00 | 99.59 | 18.54 | 142.13 |
| extra heavy standard | nr | 2.00 | 30.00 | 143.16 | 25.97 | 199.13 |
| Betula (Birch) | | | | | | |
| albosinensis | | | | | | |
| light standard | nr | 1.00 | 15.00 | 37.34 | 7.85 | 60.19 |
| standard | nr | 1.20 | 18.00 | 49.79 | 10.17 | 77.96 |
| selected standard | nr | 1.40 | 21.00 | 68.47 | 13.42 | 102.89 |
| heavy standard | nr | 1.60 | 24.00 | 99.59 | 18.54 | 142.13 |
| extra heavy standard | nr | 2.00 | 30.00 | 143.16 | 25.97 | 199.13 |

| | Unit | Labour | Hours £ | Mat'ls £ | O & P £ | Total £ |
|---|---|---|---|---|---|---|
| **Tree planting (cont'd)** | | | | | | |
| ermanii | | | | | | |
| light standard | nr | 1.00 | 15.00 | 37.34 | 7.85 | 60.19 |
| standard | nr | 1.20 | 18.00 | 49.79 | 10.17 | 77.96 |
| selected standard | nr | 1.40 | 21.00 | 68.47 | 13.42 | 102.89 |
| heavy standard | nr | 1.60 | 24.00 | 99.59 | 18.54 | 142.13 |
| extra heavy standard | nr | 2.00 | 30.00 | 143.16 | 25.97 | 199.13 |
| pendula | | | | | | |
| whip | nr | 0.30 | 4.50 | 2.16 | 1.00 | 7.66 |
| feathered | nr | 0.40 | 6.00 | 5.62 | 1.74 | 13.36 |
| light standard | nr | 1.00 | 15.00 | 11.53 | 3.98 | 30.51 |
| standard | nr | 1.20 | 18.00 | 14.02 | 4.80 | 36.82 |
| selected standard | nr | 1.40 | 21.00 | 20.24 | 6.19 | 47.43 |
| heavy standard | nr | 1.60 | 24.00 | 29.88 | 8.08 | 61.96 |
| extra heavy standard | nr | 2.00 | 30.00 | 68.47 | 14.77 | 113.24 |
| p. youngii | | | | | | |
| light standard | nr | 1.00 | 15.00 | 37.34 | 7.85 | 60.19 |
| standard | nr | 1.20 | 18.00 | 49.79 | 10.17 | 77.96 |
| selected standard | nr | 1.40 | 21.00 | 68.47 | 13.42 | 102.89 |
| heavy standard | nr | 1.60 | 24.00 | 99.59 | 18.54 | 142.13 |
| extra heavy standard | nr | 2.00 | 30.00 | 143.16 | 25.97 | 199.13 |
| Fagus (Beech) | | | | | | |
| sylvatica | | | | | | |
| whip | nr | 0.30 | 4.50 | 4.06 | 1.28 | 9.84 |
| feathered | nr | 0.40 | 6.00 | 8.71 | 2.21 | 16.92 |
| light standard | nr | 1.00 | 15.00 | 17.43 | 4.86 | 37.29 |
| standard | nr | 1.20 | 18.00 | 19.92 | 5.69 | 43.61 |
| selected standard | nr | 1.40 | 21.00 | 34.21 | 8.28 | 63.49 |
| heavy standard | nr | 1.60 | 24.00 | 56.02 | 12.00 | 92.02 |
| extra heavy standard | nr | 2.00 | 30.00 | 68.46 | 14.77 | 113.23 |

| | Unit | Labour | Hours £ | Mat'ls £ | O & P £ | Total £ |
|---|---|---|---|---|---|---|
| s. Dawyck purple | | | | | | |
| light standard | nr | 1.00 | 15.00 | 37.34 | 7.85 | 60.19 |
| standard | nr | 1.20 | 18.00 | 49.79 | 10.17 | 77.96 |
| selected standard | nr | 1.40 | 21.00 | 68.47 | 13.42 | 102.89 |
| heavy standard | nr | 1.60 | 24.00 | 99.59 | 18.54 | 142.13 |
| extra heavy standard | nr | 2.00 | 30.00 | 143.16 | 25.97 | 199.13 |
| Fraxinus (Ash) | | | | | | |
| angustifolia Raywood | | | | | | |
| whip | nr | 0.30 | 4.50 | 4.06 | 1.28 | 9.84 |
| feathered | nr | 0.40 | 6.00 | 8.71 | 2.21 | 16.92 |
| light standard | nr | 1.00 | 15.00 | 17.43 | 4.86 | 37.29 |
| standard | nr | 1.20 | 18.00 | 19.92 | 5.69 | 43.61 |
| selected standard | nr | 1.40 | 21.00 | 34.21 | 8.28 | 63.49 |
| heavy standard | nr | 1.60 | 24.00 | 56.02 | 12.00 | 92.02 |
| extra heavy standard | nr | 2.00 | 30.00 | 68.46 | 14.77 | 113.23 |
| jaspidea | | | | | | |
| whip | nr | 0.30 | 4.50 | 5.90 | 1.56 | 11.96 |
| feathered | nr | 0.40 | 6.00 | 12.45 | 2.77 | 21.22 |
| light standard | nr | 1.00 | 15.00 | 24.24 | 5.89 | 45.13 |
| standard | nr | 1.20 | 18.00 | 34.21 | 7.83 | 60.04 |
| selected standard | nr | 1.40 | 21.00 | 56.01 | 11.55 | 88.56 |
| heavy standard | nr | 1.60 | 24.00 | 68.47 | 13.87 | 106.34 |
| extra heavy standard | nr | 2.00 | 30.00 | 80.91 | 16.64 | 127.55 |
| pendula | | | | | | |
| whip | nr | 0.30 | 4.50 | 5.90 | 1.56 | 11.96 |
| feathered | nr | 0.40 | 6.00 | 12.45 | 2.77 | 21.22 |
| light standard | nr | 1.00 | 15.00 | 24.24 | 5.89 | 45.13 |
| standard | nr | 1.20 | 18.00 | 34.21 | 7.83 | 60.04 |
| selected standard | nr | 1.40 | 21.00 | 56.01 | 11.55 | 88.56 |
| heavy standard | nr | 1.60 | 24.00 | 68.47 | 13.87 | 106.34 |
| extra heavy standard | nr | 2.00 | 30.00 | 80.91 | 16.64 | 127.55 |

| | Unit | Labour | Hours £ | Mat'ls £ | O & P £ | Total £ |
|---|---|---|---|---|---|---|
| **Tree planting (cont'd)** | | | | | | |
| Prunus (Cherry) | | | | | | |
| avium | | | | | | |
| whip | nr | 0.30 | 4.50 | 2.16 | 1.00 | 7.66 |
| feathered | nr | 0.40 | 6.00 | 5.62 | 1.74 | 13.36 |
| light standard | nr | 1.00 | 15.00 | 11.53 | 3.98 | 30.51 |
| standard | nr | 1.20 | 18.00 | 14.02 | 4.80 | 36.82 |
| selected standard | nr | 1.40 | 21.00 | 20.24 | 6.19 | 47.43 |
| heavy standard | nr | 1.60 | 24.00 | 29.88 | 8.08 | 61.96 |
| extra heavy standard | nr | 2.00 | 30.00 | 68.47 | 14.77 | 113.24 |
| dulcis | | | | | | |
| whip | nr | 0.30 | 4.50 | 4.06 | 1.28 | 9.84 |
| feathered | nr | 0.40 | 6.00 | 8.71 | 2.21 | 16.92 |
| light standard | nr | 1.00 | 15.00 | 17.43 | 4.86 | 37.29 |
| standard | nr | 1.20 | 18.00 | 19.92 | 5.69 | 43.61 |
| selected standard | nr | 1.40 | 21.00 | 34.21 | 8.28 | 63.49 |
| heavy standard | nr | 1.60 | 24.00 | 56.02 | 12.00 | 92.02 |
| extra heavy standard | nr | 2.00 | 30.00 | 68.46 | 14.77 | 113.23 |
| royal burgundy | | | | | | |
| whip | nr | 0.30 | 4.50 | 5.90 | 1.56 | 11.96 |
| feathered | nr | 0.40 | 6.00 | 12.45 | 2.77 | 21.22 |
| light standard | nr | 1.00 | 15.00 | 24.24 | 5.89 | 45.13 |
| standard | nr | 1.20 | 18.00 | 34.21 | 7.83 | 60.04 |
| selected standard | nr | 1.40 | 21.00 | 56.01 | 11.55 | 88.56 |
| heavy standard | nr | 1.60 | 24.00 | 68.47 | 13.87 | 106.34 |
| extra heavy standard | nr | 2.00 | 30.00 | 80.91 | 16.64 | 127.55 |
| taihaku | | | | | | |
| whip | nr | 0.30 | 4.50 | 4.06 | 1.28 | 9.84 |
| feathered | nr | 0.40 | 6.00 | 8.71 | 2.21 | 16.92 |
| light standard | nr | 1.00 | 15.00 | 17.43 | 4.86 | 37.29 |
| standard | nr | 1.20 | 18.00 | 19.92 | 5.69 | 43.61 |

| | Unit | Labour | Hours £ | Mat'ls £ | O & P £ | Total £ |
|---|---|---|---|---|---|---|
| selected standard | nr | 1.40 | 21.00 | 34.21 | 8.28 | 63.49 |
| heavy standard | nr | 1.60 | 24.00 | 56.02 | 12.00 | 92.02 |
| extra heavy standard | nr | 2.00 | 30.00 | 68.46 | 14.77 | 113.23 |
| Quercus (Oak) | | | | | | |
| cerris | | | | | | |
| whip | nr | 0.30 | 4.50 | 5.90 | 1.56 | 11.96 |
| feathered | nr | 0.40 | 6.00 | 12.45 | 2.77 | 21.22 |
| light standard | nr | 1.00 | 15.00 | 24.24 | 5.89 | 45.13 |
| standard | nr | 1.20 | 18.00 | 34.21 | 7.83 | 60.04 |
| selected standard | nr | 1.40 | 21.00 | 56.01 | 11.55 | 88.56 |
| heavy standard | nr | 1.60 | 24.00 | 68.47 | 13.87 | 106.34 |
| extra heavy standard | nr | 2.00 | 30.00 | 80.91 | 16.64 | 127.55 |
| robur | | | | | | |
| whip | nr | 0.30 | 4.50 | 2.81 | 1.10 | 8.41 |
| feathered | nr | 0.40 | 6.00 | 6.82 | 1.92 | 14.74 |
| light standard | nr | 1.00 | 15.00 | 14.61 | 4.44 | 34.05 |
| standard | nr | 1.20 | 18.00 | 16.83 | 5.22 | 40.05 |
| selected standard | nr | 1.40 | 21.00 | 27.39 | 7.26 | 55.65 |
| heavy standard | nr | 1.60 | 24.00 | 49.79 | 11.07 | 84.86 |
| extra heavy standard | nr | 2.00 | 30.00 | 62.24 | 13.84 | 106.08 |
| rubra | | | | | | |
| whip | nr | 0.30 | 4.50 | 4.06 | 1.28 | 9.84 |
| feathered | nr | 0.40 | 6.00 | 8.71 | 2.21 | 16.92 |
| light standard | nr | 1.00 | 15.00 | 17.43 | 4.86 | 37.29 |
| standard | nr | 1.20 | 18.00 | 19.92 | 5.69 | 43.61 |
| selected standard | nr | 1.40 | 21.00 | 34.21 | 8.28 | 63.49 |
| heavy standard | nr | 1.60 | 24.00 | 56.02 | 12.00 | 92.02 |
| extra heavy standard | nr | 2.00 | 30.00 | 68.46 | 14.77 | 113.23 |

| | Unit | Labour | Hours £ | Mat'ls £ | O & P £ | Total £ |
|---|---|---|---|---|---|---|
| **Tree planting (cont'd)** | | | | | | |
| Sorbus (Whitebeam) | | | | | | |
| aria | | | | | | |
| whip | nr | 0.30 | 4.50 | 2.81 | 1.10 | 8.41 |
| feathered | nr | 0.40 | 6.00 | 6.82 | 1.92 | 14.74 |
| light standard | nr | 1.00 | 15.00 | 14.61 | 4.44 | 34.05 |
| standard | nr | 1.20 | 18.00 | 16.83 | 5.22 | 40.05 |
| selected standard | nr | 1.40 | 21.00 | 27.39 | 7.26 | 55.65 |
| heavy standard | nr | 1.60 | 24.00 | 49.79 | 11.07 | 84.86 |
| extra heavy standard | nr | 2.00 | 30.00 | 62.24 | 13.84 | 106.08 |
| aucuparia | | | | | | |
| whip | nr | 0.30 | 4.50 | 2.16 | 1.00 | 7.66 |
| feathered | nr | 0.40 | 6.00 | 5.62 | 1.74 | 13.36 |
| light standard | nr | 1.00 | 15.00 | 11.53 | 3.98 | 30.51 |
| standard | nr | 1.20 | 18.00 | 14.02 | 4.80 | 36.82 |
| selected standard | nr | 1.40 | 21.00 | 20.24 | 6.19 | 47.43 |
| heavy standard | nr | 1.60 | 24.00 | 29.88 | 8.08 | 61.96 |
| extra heavy standard | nr | 2.00 | 30.00 | 68.47 | 14.77 | 113.24 |
| hupehensis | | | | | | |
| whip | nr | 0.30 | 4.50 | 4.06 | 1.28 | 9.84 |
| feathered | nr | 0.40 | 6.00 | 8.71 | 2.21 | 16.92 |
| light standard | nr | 1.00 | 15.00 | 17.43 | 4.86 | 37.29 |
| standard | nr | 1.20 | 18.00 | 19.92 | 5.69 | 43.61 |
| selected standard | nr | 1.40 | 21.00 | 34.21 | 8.28 | 63.49 |
| heavy standard | nr | 1.60 | 24.00 | 56.02 | 12.00 | 92.02 |
| extra heavy standard | nr | 2.00 | 30.00 | 68.46 | 14.77 | 113.23 |
| sargentiana | | | | | | |
| whip | nr | 0.30 | 4.50 | 5.90 | 1.56 | 11.96 |
| feathered | nr | 0.40 | 6.00 | 12.45 | 2.77 | 21.22 |
| light standard | nr | 1.00 | 15.00 | 24.24 | 5.89 | 45.13 |
| standard | nr | 1.20 | 18.00 | 34.21 | 7.83 | 60.04 |

| | Unit | Labour | Hours £ | Mat'ls £ | O & P £ | Total £ |
|---|---|---|---|---|---|---|
| selected standard | nr | 1.40 | 21.00 | 56.01 | 11.55 | 88.56 |
| heavy standard | nr | 1.60 | 24.00 | 68.47 | 13.87 | 106.34 |
| extra heavy standard | nr | 2.00 | 30.00 | 80.91 | 16.64 | 127.55 |
| torminalis | | | | | | |
| whip | nr | 0.30 | 4.50 | 5.90 | 1.56 | 11.96 |
| feathered | nr | 0.40 | 6.00 | 12.45 | 2.77 | 21.22 |
| light standard | nr | 1.00 | 15.00 | 24.24 | 5.89 | 45.13 |
| standard | nr | 1.20 | 18.00 | 34.21 | 7.83 | 60.04 |
| selected standard | nr | 1.40 | 21.00 | 56.01 | 11.55 | 88.56 |
| heavy standard | nr | 1.60 | 24.00 | 68.47 | 13.87 | 106.34 |
| extra heavy standard | nr | 2.00 | 30.00 | 80.91 | 16.64 | 127.55 |
| Tilia (Lime) | | | | | | |
| cordata | | | | | | |
| whip | nr | 0.30 | 4.50 | 2.81 | 1.10 | 8.41 |
| feathered | nr | 0.40 | 6.00 | 6.82 | 1.92 | 14.74 |
| light standard | nr | 1.00 | 15.00 | 14.61 | 4.44 | 34.05 |
| standard | nr | 1.20 | 18.00 | 16.83 | 5.22 | 40.05 |
| selected standard | nr | 1.40 | 21.00 | 27.39 | 7.26 | 55.65 |
| heavy standard | nr | 1.60 | 24.00 | 49.79 | 11.07 | 84.86 |
| extra heavy standard | nr | 2.00 | 30.00 | 62.24 | 13.84 | 106.08 |
| petiolaris | | | | | | |
| whip | nr | 0.30 | 4.50 | 4.06 | 1.28 | 9.84 |
| feathered | nr | 0.40 | 6.00 | 8.71 | 2.21 | 16.92 |
| light standard | nr | 1.00 | 15.00 | 17.43 | 4.86 | 37.29 |
| standard | nr | 1.20 | 18.00 | 19.92 | 5.69 | 43.61 |
| selected standard | nr | 1.40 | 21.00 | 34.21 | 8.28 | 63.49 |
| heavy standard | nr | 1.60 | 24.00 | 56.02 | 12.00 | 92.02 |
| extra heavy standard | nr | 2.00 | 30.00 | 68.46 | 14.77 | 113.23 |

| | Unit | Labour | Hours £ | Mat'ls £ | O & P £ | Total £ |
|---|---|---|---|---|---|---|
| **Tree planting (cont'd)** | | | | | | |
| p. rubra | | | | | | |
| whip | nr | 0.30 | 4.50 | 4.06 | 1.28 | 9.84 |
| feathered | nr | 0.40 | 6.00 | 8.71 | 2.21 | 16.92 |
| light standard | nr | 1.00 | 15.00 | 17.43 | 4.86 | 37.29 |
| standard | nr | 1.20 | 18.00 | 19.92 | 5.69 | 43.61 |
| selected standard | nr | 1.40 | 21.00 | 34.21 | 8.28 | 63.49 |
| heavy standard | nr | 1.60 | 24.00 | 56.02 | 12.00 | 92.02 |
| extra heavy standard | nr | 2.00 | 30.00 | 68.46 | 14.77 | 113.23 |
| **CONTAINER GROWN TREES** | | | | | | |
| **Excavate tree pit, fork bottom, plant tree, backfill with excavated material including organic manure (30% of soil by volume), water and surround with peat** | | | | | | |
| Acer (Maple) | | | | | | |
| campestre carnival, size 6-8cm, pot 25 litre | nr | 1.00 | 15.00 | 34.92 | 7.49 | 57.41 |
| c. Postelense, size 6-8cm, pot 25 litre | nr | 1.00 | 15.00 | 34.92 | 7.49 | 57.41 |
| negundo elegans, size 6-8cm pot 25 litre | nr | 1.00 | 15.00 | 32.43 | 7.11 | 54.54 |
| p. Deborah, size 6-8cm, pot 25 litre | nr | 1.00 | 15.00 | 29.93 | 6.74 | 51.67 |
| p. Deborah, size 10-12cm, pot 45 litre | nr | 1.40 | 21.00 | 69.85 | 13.63 | 104.48 |

| | Unit | Labour | Hours £ | Mat'ls £ | O & P £ | Total £ |
|---|---|---|---|---|---|---|
| p. Drummondii, size 6-8cm, pot 25 litre | nr | 1.00 | 15.00 | 29.93 | 6.74 | 51.67 |
| p. Drummondii, size 10-12cm, pot 45 litre | nr | 1.40 | 21.00 | 69.85 | 13.63 | 104.48 |
| Aesculus (Chestnut) | | | | | | |
| x carnea Briotii, size 6-8cm, pot 25 litre | nr | 1.00 | 15.00 | 29.93 | 6.74 | 51.67 |
| x carnea Briotii, size 10-12cm, pot 45 litre | nr | 1.40 | 21.00 | 69.85 | 13.63 | 104.48 |
| Alnus | | | | | | |
| incarna Aurea, size 6-8cm, pot 25 litre | nr | 1.00 | 15.00 | 32.43 | 7.11 | 54.54 |
| incarna Aurea, size 10-12cm, pot 45 litre | nr | 1.40 | 21.00 | 89.80 | 16.62 | 127.42 |
| Betula (Birch) | | | | | | |
| albosinensis China Ruby size 6-8cm, pot 25 litre | nr | 1.00 | 15.00 | 34.92 | 7.49 | 57.41 |
| nigra, size 10-12cm, pot 45 litre | nr | 1.40 | 21.00 | 85.46 | 15.97 | 122.43 |
| pendula, size 6-8cm, pot 25 litre | nr | 1.00 | 15.00 | 23.70 | 5.81 | 44.51 |
| pendula, size 10-12cm, pot 45 litre | nr | 1.40 | 21.00 | 66.10 | 13.07 | 100.17 |
| Laciniata, size 6-8cm, pot 25 litre | nr | 1.00 | 15.00 | 34.92 | 7.49 | 57.41 |
| Laciniata, size 10-12cm, pot 45 litre | nr | 1.40 | 21.00 | 86.06 | 16.06 | 123.12 |

| | Unit | Labour | Hours £ | Mat'ls £ | O & P £ | Total £ |
|---|---|---|---|---|---|---|
| **Tree planting (cont'd)** | | | | | | |
| p. Youngii, size 6-8cm, pot 25 litre | nr | 1.00 | 15.00 | 34.92 | 7.49 | 57.41 |
| p. Youngii, size 10-12cm, pot 45 litre | nr | 1.40 | 21.00 | 84.81 | 15.87 | 121.68 |
| utilis var. jacquentmontii Doorenbos, size 6-8cm, pot 25 litre | nr | 1.00 | 15.00 | 34.92 | 7.49 | 57.41 |
| utilis var. jacquentmontii Doorenbos, size 10-12cm, pot 45 litre | nr | 1.40 | 21.00 | 84.81 | 15.87 | 121.68 |
| Fagus (Beech) | | | | | | |
| sylvatica, size 6-8cm, pot 25 litre | nr | 1.00 | 15.00 | 29.93 | 6.74 | 51.67 |
| sylvatica, size 10-12cm, pot 45 litre | nr | 1.40 | 21.00 | 110.00 | 19.65 | 150.65 |
| s. Altropurperea, size 6-8cm, pot 25 litre | nr | 1.00 | 15.00 | 34.92 | 7.49 | 57.41 |
| s. Altropurperea, size 10-12cm, pot 45 litre | nr | 1.40 | 21.00 | 110.00 | 19.65 | 150.65 |
| Fraxinus (Ash) | | | | | | |
| Augustifolia Raywood, size 6-8cm, pot 25 litre | nr | 1.00 | 15.00 | 29.93 | 6.74 | 51.67 |
| Augustifolia Raywood, size 10-12cm, pot 45 litre | nr | 1.40 | 21.00 | 67.35 | 13.25 | 101.60 |
| excelsior Jaspidea, size 6-8cm, pot 25 litre | nr | 1.00 | 15.00 | 34.92 | 7.49 | 57.41 |
| excelsior Jaspidea, size 10-12cm, pot 45 litre | nr | 1.40 | 21.00 | 77.33 | 14.75 | 113.08 |

| | Unit | Labour | Hours £ | Mat'ls £ | O & P £ | Total £ |
|---|---|---|---|---|---|---|
| Prunus (Cherry) | | | | | | |
| Accolade, size 6-8cm, pot 25 litre | nr | 1.00 | 15.00 | 29.93 | 6.74 | 51.67 |
| avium Plena, size 6-8cm, pot 25 litre | nr | 1.00 | 15.00 | 29.93 | 6.74 | 51.67 |
| avium Plena, size 10-12cm, pot 45 litre | nr | 1.40 | 21.00 | 67.35 | 13.25 | 101.60 |
| Fragrant Cloud, size 6-8cm, pot 25 litre | nr | 1.00 | 15.00 | 29.93 | 6.74 | 51.67 |
| Kanzan, size 6-8cm, pot 25 litre | nr | 1.00 | 15.00 | 29.93 | 6.74 | 51.67 |
| Kanzan, size 10-12cm, pot 45 litre | nr | 1.40 | 21.00 | 67.35 | 13.25 | 101.60 |
| Pandora, size 6-8cm, pot 25 litre | nr | 1.00 | 15.00 | 29.93 | 6.74 | 51.67 |
| Royal Burgundy, size 6-8cm, pot 25 litre | nr | 1.00 | 15.00 | 34.92 | 7.49 | 57.41 |
| sargentii, size 6-8cm, pot 25 litre | nr | 1.00 | 15.00 | 29.93 | 6.74 | 51.67 |
| serrula, size 6-8cm, pot 25 litre | nr | 1.00 | 15.00 | 29.93 | 6.74 | 51.67 |
| x s. Autumnalis Rosea, size 6-8cm, pot 25 litre | nr | 1.00 | 15.00 | 29.93 | 6.74 | 51.67 |
| x s. Pendula Rubra, size 6-8cm, pot 25 litre | nr | 1.00 | 15.00 | 29.93 | 6.74 | 51.67 |
| Umineko, size 6-8cm, pot 25 litre | nr | 1.00 | 15.00 | 29.93 | 6.74 | 51.67 |
| Umineko, size 10-12cm, pot 45 litre | nr | 1.40 | 21.00 | 29.93 | 7.64 | 58.57 |
| x yeodensis, size 6-8cm, pot 25 litre | nr | 1.00 | 15.00 | 29.93 | 6.74 | 51.67 |
| x y Shidare-yoshino, size 6-8cm, pot 25 litre | nr | 1.00 | 15.00 | 29.93 | 6.74 | 51.67 |

| | Unit | Labour | Hours £ | Mat'ls £ | O & P £ | Total £ |
|---|---|---|---|---|---|---|
| **Tree planting (cont'd)** | | | | | | |
| Quercus (Oak) | | | | | | |
| ilex, size 6-8cm, pot 25 litre | nr | 1.00 | 15.00 | 34.97 | 7.50 | 57.47 |
| ilex, size 10-12cm, pot 45 litre | nr | 1.40 | 21.00 | 78.57 | 14.94 | 114.51 |
| robur f. fastigiata, size 10-12cm, pot 45 litre | nr | 1.40 | 21.00 | 112.74 | 20.06 | 153.80 |
| Sorbus (Whitebeam) | | | | | | |
| aria Lutescens, size 6-8cm, pot 25 litre | nr | 1.00 | 15.00 | 29.93 | 6.74 | 51.67 |
| aria Lutescens, size 10-12cm, pot 45 litre | nr | 1.40 | 21.00 | 67.35 | 13.25 | 101.60 |
| majestica, size 6-8cm, pot 25 litre | nr | 1.00 | 15.00 | 29.93 | 6.74 | 51.67 |
| majestica, size 10-12cm, pot 45 litre | nr | 1.40 | 21.00 | 67.35 | 13.25 | 101.60 |
| a. var. xanthocarpa, size 6-8cm, pot 25 litre | nr | 1.00 | 15.00 | 29.93 | 6.74 | 51.67 |
| commixta Embley, size 6-8cm, pot 25 litre | nr | 1.00 | 15.00 | 29.93 | 6.74 | 51.67 |
| hupehensis, size 6-8cm, pot 25 litre | nr | 1.00 | 15.00 | 29.93 | 6.74 | 51.67 |
| hupehensis, size 10-12cm, pot 45 litre | nr | 1.40 | 21.00 | 77.33 | 14.75 | 113.08 |
| Leonard Messel, size 6-8cm, pot 25 litre | nr | 1.00 | 15.00 | 29.93 | 6.74 | 51.67 |
| sargentiana, size 6-8cm, pot 25 litre | nr | 1.00 | 15.00 | 29.93 | 6.74 | 51.67 |
| Sunshine, size 6-8cm, pot 25 litre | nr | 1.00 | 15.00 | 29.93 | 6.74 | 51.67 |

| | Unit | Labour | Hours £ | Mat'ls £ | O & P £ | Total £ |
|---|---|---|---|---|---|---|
| **CONIFERS** | | | | | | |
| **Excavate tree pit, fork bottom, plant conifer, backfill with excavated material including organic manure (30% of soil by volume), water and surround with peat** | | | | | | |
| Abies (Fir) | | | | | | |
| concolor, height 60cm, pot 3 litre | nr | 0.30 | 4.50 | 4.72 | 1.38 | 10.60 |
| Cedrus (Cedar) | | | | | | |
| antlantica, height 175cm, pot 35 litre | nr | 1.00 | 15.00 | 68.60 | 12.54 | 96.14 |
| antlantica, height 200cm, pot 50 litre | nr | 1.00 | 15.00 | 99.78 | 17.22 | 132.00 |
| antlantica, height 250cm, pot 75 litre | nr | 1.00 | 15.00 | 149.67 | 24.70 | 189.37 |
| deodara, height 175cm, pot 30 litre | nr | 1.00 | 15.00 | 56.13 | 10.67 | 81.80 |
| deodara, height 200cm, pot 35 litre | nr | 1.00 | 15.00 | 81.07 | 14.41 | 110.48 |
| deodara, height 250cm, pot 50 litre | nr | 1.00 | 15.00 | 99.78 | 17.22 | 132.00 |
| libani, height 175cm, pot 30 litre | nr | 1.00 | 15.00 | 56.13 | 10.67 | 81.80 |
| libani, height 200cm, pot 35 litre | nr | 1.00 | 15.00 | 81.07 | 14.41 | 110.48 |
| libani, height 250cm, pot 50 litre | nr | 1.00 | 15.00 | 99.78 | 17.22 | 132.00 |

| | Unit | Labour | Hours £ | Mat'ls £ | O & P £ | Total £ |
|---|---|---|---|---|---|---|
| **Tree planting (cont'd)** | | | | | | |
| Chamaecyparis (Cypress) | | | | | | |
| lawsoniana, height 80cm, pot 3 litre | nr | 0.30 | 4.50 | 4.72 | 1.38 | 10.60 |
| l. Ellwoodii, height 40cm, pot 2 litre | nr | 0.30 | 4.50 | 4.72 | 1.38 | 10.60 |
| l. Fletcherii, height 80cm, pot 3 litre | nr | 0.30 | 4.50 | 4.72 | 1.38 | 10.60 |
| l. stardust, height 80cm, pot 3 litre | nr | 0.30 | 4.50 | 4.72 | 1.38 | 10.60 |
| Crytopmeria (Cedar) | | | | | | |
| japonica, height 60cm, pot 3 litre | nr | 0.30 | 4.50 | 5.86 | 1.55 | 11.91 |
| leylandii, height 80cm, pot 3 litre | nr | 0.30 | 4.50 | 3.74 | 1.24 | 9.48 |
| leylandii, height 100cm, pot 5 litre | nr | 0.30 | 4.50 | 7.16 | 1.75 | 13.41 |
| leylandii, height 125cm, pot 10 litre | nr | 0.30 | 4.50 | 9.98 | 2.17 | 16.65 |
| leylandii, height 150cm, pot 10 litre | nr | 0.30 | 4.50 | 12.47 | 2.55 | 19.52 |
| leylandii, height 175cm, pot 13 litre | nr | 0.30 | 4.50 | 18.71 | 3.48 | 26.69 |
| Cupressus (Cypress) | | | | | | |
| macrocarpa Goldcrest, height 125cm, pot 10 litre | nr | 0.30 | 4.50 | 29.93 | 5.16 | 39.59 |
| macrocarpa Goldcrest, height 150cm, pot 15 litre | nr | 0.30 | 4.50 | 49.86 | 8.15 | 62.51 |

| | Unit | Labour | Hours £ | Mat'ls £ | O & P £ | Total £ |
|---|---|---|---|---|---|---|
| macrocarpa Goldcrest, height 200cm, pot 25 litre | nr | 0.30 | 4.50 | 74.83 | 11.90 | 91.23 |
| macrocarpa Goldcrest, height 250cm, pot 30 litre | nr | 0.30 | 4.50 | 99.78 | 15.64 | 119.92 |
| Larix (Larch) | | | | | | |
| decidua, height 80cm, pot 3 litre | nr | 0.30 | 4.50 | 4.77 | 1.39 | 10.66 |
| x eurolepis, height 80cm, pot 3 litre | nr | 0.30 | 4.50 | 4.77 | 1.39 | 10.66 |
| kaempferi, height 80cm, pot 3 litre | nr | 0.30 | 4.50 | 4.77 | 1.39 | 10.66 |
| Pinus (Pine) | | | | | | |
| nigra subsp laricio, height 80cm, pot 3 litre | nr | 0.30 | 4.50 | 4.33 | 1.32 | 10.15 |
| nigra subsp laricio, height 100cm, pot 10 litre | nr | 0.30 | 4.50 | 12.47 | 2.55 | 19.52 |
| nigra subsp laricio, height 125cm, pot 10 litre | nr | 0.30 | 4.50 | 13.72 | 2.73 | 20.95 |
| mugo, height 30cm, pot 2 litre | nr | 0.30 | 4.50 | 5.64 | 1.52 | 11.66 |
| mugo, height 50cm, pot 10 litre | nr | 0.30 | 4.50 | 18.71 | 3.48 | 26.69 |

| | Unit | Labour | Hours £ | Mat'ls £ | O & P £ | Total £ |
|---|---|---|---|---|---|---|
| Taxus (Yew) | | | | | | |
| baccata, height 60cm, pot 3 litre | nr | 0.30 | 4.50 | 5.64 | 1.52 | 11.66 |
| baccata, height 80cm, pot 5 litre | nr | 0.30 | 4.50 | 8.73 | 1.98 | 15.21 |
| baccata, height 100cm, pot 10 litre | nr | 0.30 | 4.50 | 18.71 | 3.48 | 26.69 |
| baccata, height 125cm, pot 20 litre | nr | 0.30 | 4.50 | 37.47 | 6.30 | 48.27 |

**SHRUBS**

**The following descriptions refer to dimensions and sizes of shrubs as as set out below.**

Dwarf: 20 to 60cm high
Small: 60 to 120cm high
Medium: 120 to 200cm high
Large: over 200cm high

**Form planting hole in cultivated area, place shrub in hole, backfill, water and surround with peat**

| | Unit | Labour | Hours £ | Mat'ls £ | O & P £ | Total £ |
|---|---|---|---|---|---|---|
| Acer (Maple) | | | | | | |
| campestre, dwarf, pot 3 litre | nr | 0.20 | 3.00 | 2.50 | 0.83 | 6.33 |
| campestre, medium, pot 10 litre | nr | 0.40 | 6.00 | 12.47 | 2.77 | 21.24 |
| p. f. atropurpureum, dwarf, pot 3 litre | nr | 0.20 | 3.00 | 13.72 | 2.51 | 19.23 |

| | Unit | Labour | Hours £ | Mat'ls £ | O & P £ | Total £ |
|---|---|---|---|---|---|---|
| p. f. atropurpureum, small, pot 10 litre | nr | 0.30 | 4.50 | 31.18 | 5.35 | 41.03 |
| p. Bloodgood, small, pot 3 litre | nr | 0.30 | 4.50 | 13.72 | 2.73 | 20.95 |
| Sango-kaku, dwarf, pot 3 litre | nr | 0.20 | 3.00 | 13.72 | 2.51 | 19.23 |
| Sango-kaku, small, pot 3 litre | nr | 0.30 | 4.50 | 31.18 | 5.35 | 41.03 |
| shirawanum, dwarf pot 3 litre | nr | 0.20 | 3.00 | 13.72 | 2.51 | 19.23 |
| Buddleja (Buddleia) | | | | | | |
| davidii Black Knight, dwarf, pot 3 litre | nr | 0.20 | 3.00 | 4.33 | 1.10 | 8.43 |
| davidii Black Knight, small, pot 10 litre | nr | 0.30 | 4.50 | 13.72 | 2.73 | 20.95 |
| d. Harlequin, dwarf, pot 3 litre | nr | 0.20 | 3.00 | 4.33 | 1.10 | 8.43 |
| d. Harlequin, small, pot 10 litre | nr | 0.30 | 4.50 | 13.72 | 2.73 | 20.95 |
| d. Royal Red, dwarf, pot 3 litre | nr | 0.20 | 3.00 | 4.33 | 1.10 | 8.43 |
| d. Royal Red, small, pot 10 litre | nr | 0.30 | 4.50 | 13.72 | 2.73 | 20.95 |
| d. White Profusion, dwarf, pot 3 litre | nr | 0.20 | 3.00 | 4.33 | 1.10 | 8.43 |
| d. White Profusion, small, pot 10 litre | nr | 0.30 | 4.50 | 13.72 | 2.73 | 20.95 |
| Pink Delight, dwarf, pot 3 litre | nr | 0.20 | 3.00 | 4.33 | 1.10 | 8.43 |
| Pink Delight, small, pot 10 litre | nr | 0.30 | 4.50 | 13.72 | 2.73 | 20.95 |

| | Unit | Labour | Hours £ | Mat'ls £ | O & P £ | Total £ |
|---|---|---|---|---|---|---|
| Camellia (Camellia) | | | | | | |
| japonica Adolphe Audusson, dwarf, pot 3 litre | nr | 0.20 | 3.00 | 6.24 | 1.39 | 10.63 |
| japonica Adolphe Audusson, small, pot 3 litre | nr | 0.30 | 4.50 | 8.13 | 1.89 | 14.52 |
| japonica Adolphe Audusson, medium, pot 10 litre | nr | 0.40 | 6.00 | 17.46 | 3.52 | 26.98 |
| j. Lady Vansittart, dwarf, pot 3 litre | nr | 0.20 | 3.00 | 6.24 | 1.39 | 10.63 |
| j. Lady Vansittart, small, pot 3 litre | nr | 0.30 | 4.50 | 8.13 | 1.89 | 14.52 |
| j. Lady Vansittart, medium, pot 10 litre | nr | 0.40 | 6.00 | 17.46 | 3.52 | 26.98 |
| Leonard Messel, dwarf, pot 3 litre | nr | 0.20 | 3.00 | 6.24 | 1.39 | 10.63 |
| Leonard Messel, small, pot 3 litre | nr | 0.30 | 4.50 | 8.13 | 1.89 | 14.52 |
| Leonard Messel, medium, pot 10 litre | nr | 0.40 | 6.00 | 17.46 | 3.52 | 26.98 |
| x w. Debbie, dwarf, pot 3 litre | nr | 0.20 | 3.00 | 6.24 | 1.39 | 10.63 |
| Leonard Messel, small, pot 3 litre | nr | 0.30 | 4.50 | 8.13 | 1.89 | 14.52 |
| Leonard Messel, medium, pot 10 litre | nr | 0.40 | 6.00 | 17.46 | 3.52 | 26.98 |
| Cytisus (Broom) | | | | | | |
| battandieri, dwarf, pot 3 litre | nr | 0.20 | 3.00 | 8.13 | 1.67 | 12.80 |
| battandieri, small, pot 10 litre | nr | 0.30 | 4.50 | 17.46 | 3.29 | 25.25 |

| | Unit | Labour | Hours £ | Mat'ls £ | O & P £ | Total £ |
|---|---|---|---|---|---|---|
| burkwoodii, dwarf, pot 3 litre | nr | 0.20 | 3.00 | 4.33 | 1.10 | 8.43 |
| burkwoodii, small, pot 10 litre | nr | 0.30 | 4.50 | 13.72 | 2.73 | 20.95 |
| Fulgens, dwarf, pot 3 litre | nr | 0.20 | 3.00 | 4.33 | 1.10 | 8.43 |
| Fulgens, small, pot 10 litre | nr | 0.30 | 4.50 | 13.72 | 2.73 | 20.95 |
| Goldfinch, dwarf, pot 3 litre | nr | 0.20 | 3.00 | 4.33 | 1.10 | 8.43 |
| Goldfinch, small, pot 10 litre | nr | 0.30 | 4.50 | 13.72 | 2.73 | 20.95 |
| Lena, dwarf, pot 3 litre | nr | 0.20 | 3.00 | 4.33 | 1.10 | 8.43 |
| Lena, small, pot 10 litre | nr | 0.30 | 4.50 | 13.72 | 2.73 | 20.95 |
| x p. Allgold, dwarf, pot 3 litre | nr | 0.20 | 3.00 | 4.33 | 1.10 | 8.43 |
| x p. Allgold, small, pot 10 litre | nr | 0.30 | 4.50 | 13.72 | 2.73 | 20.95 |
| Zeelandia, dwarf, pot 3 litre | nr | 0.20 | 3.00 | 4.33 | 1.10 | 8.43 |
| Fagus (Beech) | | | | | | |
| sylvatica, dwarf, pot 3 litre | nr | 0.20 | 3.00 | 2.50 | 0.83 | 6.33 |
| sylvatica, dwarf, pot 10 litre | nr | 0.30 | 4.50 | 12.47 | 2.55 | 19.52 |
| Hydrangea | | | | | | |
| arborescens Annabelle, dwarf, pot 3 litre | nr | 0.20 | 3.00 | 4.33 | 1.10 | 8.43 |
| arborescens Annabelle, small, pot 10 litre | nr | 0.30 | 4.50 | 13.72 | 2.73 | 20.95 |

| | Unit | Labour | Hours £ | Mat'ls £ | O & P £ | Total £ |
|---|---|---|---|---|---|---|
| **Shrub planting (cont'd)** | | | | | | |
| a Villosa Group, dwarf, pot 3 litre | nr | 0.20 | 3.00 | 6.24 | 1.39 | 10.63 |
| a Villosa Group, small, pot 10 litre | nr | 0.30 | 4.50 | 13.72 | 2.73 | 20.95 |
| m Mariesii Grandiflora, dwarf, pot 3 litre | nr | 0.20 | 3.00 | 4.33 | 1.10 | 8.43 |
| m Mariesii Grandiflora, small, pot 10 litre | nr | 0.30 | 4.50 | 13.72 | 2.73 | 20.95 |
| m Mowe, dwarf, pot 3 litre | nr | 0.20 | 3.00 | 4.33 | 1.10 | 8.43 |
| m Mowe, dwarf, pot 10 litre | nr | 0.30 | 4.50 | 13.72 | 2.73 | 20.95 |
| Preziosa, dwarf, pot 3 litre | nr | 0.20 | 3.00 | 4.33 | 1.10 | 8.43 |
| Preziosa, dwarf, pot 10 litre | nr | 0.30 | 4.50 | 13.72 | 2.73 | 20.95 |
| q. Snow Queen, dwarf, pot 3 litre | nr | 0.20 | 3.00 | 6.24 | 1.39 | 10.63 |
| q. Snow Queen, small, pot 10 litre | nr | 0.30 | 4.50 | 19.38 | 3.58 | 27.46 |
| s. Miranda, dwarf, pot 3 litre | nr | 0.20 | 3.00 | 6.24 | 1.39 | 10.63 |
| Lavandula (Lavender) | | | | | | |
| augustifolia, dwarf, pot 2 litre | nr | 0.20 | 3.00 | 2.49 | 0.82 | 6.31 |
| a. Munstead, dwarf, pot 2 litre | nr | 0.20 | 3.00 | 2.49 | 0.82 | 6.31 |
| a. Twickel Purple, dwarf, pot 2 litre | nr | 0.20 | 3.00 | 2.49 | 0.82 | 6.31 |
| stoechas, dwarf, pot 2 litre | nr | 0.20 | 3.00 | 2.49 | 0.82 | 6.31 |

| | Unit | Labour | Hours<br>£ | Mat'ls<br>£ | O & P<br>£ | Total<br>£ |
|---|---|---|---|---|---|---|
| Ligustrum (Privet) | | | | | | |
| japonicum, dwarf, pot 3 litre | nr | 0.20 | 3.00 | 4.99 | 1.20 | 9.19 |
| japonicum, small, pot 10 litre | nr | 0.30 | 4.50 | 13.72 | 2.73 | 20.95 |
| ovalifolium, dwarf, pot 3 litre | nr | 0.20 | 3.00 | 2.49 | 0.82 | 6.31 |
| ovalifolium, small, pot 10 litre | nr | 0.30 | 4.50 | 12.47 | 2.55 | 19.52 |
| o. Aureum, dwarf, pot 3 litre | nr | 0.20 | 3.00 | 3.74 | 1.01 | 7.75 |
| o. Aureum, small, pot 10 litre | nr | 0.30 | 4.50 | 13.72 | 2.73 | 20.95 |
| vulgare, dwarf, pot 3 litre | nr | 0.20 | 3.00 | 2.49 | 0.82 | 6.31 |
| vulgare, small, pot 10 litre | nr | 0.30 | 4.50 | 12.47 | 2.55 | 19.52 |
| Lonicera (Honeysuckle) | | | | | | |
| fragrantissima, dwarf, pot 3 litre | nr | 0.20 | 3.00 | 6.24 | 1.39 | 10.63 |
| fragrantissima, small, pot 10 litre | nr | 0.30 | 4.50 | 13.72 | 2.73 | 20.95 |
| nitida, dwarf, pot 3 litre | nr | 0.20 | 3.00 | 2.49 | 0.82 | 6.31 |
| nitida, small, pot 10 litre | nr | 0.30 | 4.50 | 13.72 | 2.73 | 20.95 |
| n. Baggesen's Gold, dwarf, pot 3 litre | nr | 0.20 | 3.00 | 3.74 | 1.01 | 7.75 |
| n. Baggesen's Gold, small, pot 10 litre | nr | 0.30 | 4.50 | 13.72 | 2.73 | 20.95 |
| pileata, dwarf, pot 3 litre | nr | 0.20 | 3.00 | 2.49 | 0.82 | 6.31 |

| | Unit | Labour | Hours £ | Mat'ls £ | O & P £ | Total £ |
|---|---|---|---|---|---|---|
| **Shrub planting (cont'd)** | | | | | | |
| pileata, small, pot 10 litre | nr | 0.20 | 3.00 | 2.49 | 0.82 | 6.31 |
| syringantha, dwarf, pot 3 litre | nr | 0.20 | 3.00 | 6.24 | 1.39 | 10.63 |
| Magnolia | | | | | | |
| x brooklynensis, dwarf, pot 3 litre | nr | 0.20 | 3.00 | 11.22 | 2.13 | 16.35 |
| g. Galisonniere, dwarf, pot 3 litre | nr | 0.20 | 3.00 | 11.22 | 2.13 | 16.35 |
| g. Galisonniere, small, pot 10 litre | nr | 0.30 | 4.50 | 28.04 | 4.88 | 37.42 |
| Heaven Scent, dwarf, pot 3 litre | nr | 0.20 | 3.00 | 11.22 | 2.13 | 16.35 |
| Heaven Scent, small, pot 10 litre | nr | 0.30 | 4.50 | 28.04 | 4.88 | 37.42 |
| liliflora Nigra, dwarf, pot 3 litre | nr | 0.20 | 3.00 | 11.22 | 2.13 | 16.35 |
| liliflora Nigra, small, pot 10 litre | nr | 0.30 | 4.50 | 28.04 | 4.88 | 37.42 |
| x soulangeana, dwarf, pot 3 litre | nr | 0.20 | 3.00 | 11.22 | 2.13 | 16.35 |
| x soulangeana, small, pot 10 litre | nr | 0.30 | 4.50 | 28.04 | 4.88 | 37.42 |
| stellata, dwarf, pot 3 litre | nr | 0.20 | 3.00 | 11.22 | 2.13 | 16.35 |
| stellata, small, pot 10 litre | nr | 0.30 | 4.50 | 28.04 | 4.88 | 37.42 |
| s. Royal Star, dwarf, pot 3 litre | nr | 0.20 | 3.00 | 11.22 | 2.13 | 16.35 |
| Sundance, dwarf, pot 3 litre | nr | 0.20 | 3.00 | 11.22 | 2.13 | 16.35 |

| | Unit | Labour | Hours £ | Mat'ls £ | O & P £ | Total £ |
|---|---|---|---|---|---|---|
| Prunus (Cherry) | | | | | | |
| cerasifera Nigra, dwarf, pot 3 litre | nr | 0.20 | 3.00 | 6.24 | 1.39 | 10.63 |
| cerasifera Nigra, small, pot 10 litre | nr | 0.30 | 4.50 | 13.72 | 2.73 | 20.95 |
| laurocerasus, dwarf, pot 3 litre | nr | 0.20 | 3.00 | 2.49 | 0.82 | 6.31 |
| laurocerasus, small, pot 10 litre | nr | 0.30 | 4.50 | 13.72 | 2.73 | 20.95 |
| l. Cherry Brandy, dwarf, pot 3 litre | nr | 0.20 | 3.00 | 4.99 | 1.20 | 9.19 |
| l. Cherry Brandy, small, pot 10 litre | nr | 0.30 | 4.50 | 13.72 | 2.73 | 20.95 |
| l. Mount Vernon, dwarf, pot 3 litre | nr | 0.20 | 3.00 | 4.99 | 1.20 | 9.19 |
| l. Mount Vernon, small, pot 10 litre | nr | 0.30 | 4.50 | 13.72 | 2.73 | 20.95 |
| l. Schipkaensis, dwarf, pot 3 litre | nr | 0.20 | 3.00 | 3.74 | 1.01 | 7.75 |
| l. Schipkaensis, small, pot 10 litre | nr | 0.30 | 4.50 | 13.72 | 2.73 | 20.95 |
| l. Zabeliana, dwarf, pot 3 litre | nr | 0.20 | 3.00 | 3.74 | 1.01 | 7.75 |
| l. Zabeliana, small, pot 10 litre | nr | 0.30 | 4.50 | 13.72 | 2.73 | 20.95 |
| lusitanica, dwarf, pot 3 litre | nr | 0.20 | 3.00 | 3.74 | 1.01 | 7.75 |
| lusitanica, small, pot 10 litre | nr | 0.30 | 4.50 | 13.72 | 2.73 | 20.95 |
| l. Variegata, dwarf, pot 3 litre | nr | 0.20 | 3.00 | 3.74 | 1.01 | 7.75 |
| l. Variegata, small, pot 10 litre | nr | 0.30 | 4.50 | 13.72 | 2.73 | 20.95 |

| | Unit | Labour | Hours £ | Mat'ls £ | O & P £ | Total £ |
|---|---|---|---|---|---|---|
| **Shrub planting (cont'd)** | | | | | | |
| spinosa, dwarf, pot 3 litre | nr | 0.20 | 3.00 | 2.49 | 0.82 | 6.31 |
| Pyracantha (Firethorn) | | | | | | |
| Red column, dwarf, pot 2 litre | nr | 0.20 | 3.00 | 3.15 | 0.92 | 7.07 |
| Red column, small, pot 10 litre | nr | 0.30 | 4.50 | 13.72 | 2.73 | 20.95 |
| Mohave, dwarf, pot 2 litre | nr | 0.20 | 3.00 | 3.15 | 0.92 | 7.07 |
| Mohave, dwarf, pot 10 litre | nr | 0.30 | 4.50 | 13.72 | 2.73 | 20.95 |
| Orange Glow, dwarf, pot 2 litre | nr | 0.20 | 3.00 | 3.58 | 0.99 | 7.57 |
| Orange Glow, small, pot 10 litre | nr | 0.30 | 4.50 | 13.72 | 2.73 | 20.95 |
| Soleil d'Or, dwarf, pot 2 litre | nr | 0.20 | 3.00 | 3.58 | 0.99 | 7.57 |
| Soleil d'Or, small, pot 10 litre | nr | 0.30 | 4.50 | 13.72 | 2.73 | 20.95 |
| Sparkler, dwarf, pot 2 litre | nr | 0.20 | 3.00 | 3.15 | 0.92 | 7.07 |
| Teton, dwarf, pot 2 litre | nr | 0.20 | 3.00 | 3.15 | 0.92 | 7.07 |
| Salix (Willow) | | | | | | |
| hastata Wehrhahnii, dwarf, pot 3 litre | nr | 0.20 | 3.00 | 6.24 | 1.39 | 10.63 |
| hastata Wehrhahnii, small, pot 10 litre | nr | 0.30 | 4.50 | 13.72 | 2.73 | 20.95 |
| helvetica, dwarf, pot 2 litre | nr | 0.20 | 3.00 | 6.24 | 1.39 | 10.63 |

| | Unit | Labour | Hours £ | Mat'ls £ | O & P £ | Total £ |
|---|---|---|---|---|---|---|
| helvetica, small, pot 10 litre | nr | 0.30 | 4.50 | 13.72 | 2.73 | 20.95 |
| viminalis, dwarf, pot 3 litre | nr | 0.20 | 3.00 | 2.49 | 0.82 | 6.31 |
| viminalis, small, pot 10 litre | nr | 0.30 | 4.50 | 12.47 | 2.55 | 19.52 |
| Syringa (Lilac) | | | | | | |
| meyeri var.spontanea Palibin, dwarf, pot 3 litre | nr | 0.20 | 3.00 | 7.16 | 1.52 | 11.68 |
| meyeri var.spontanea Palibin, small, pot 10 litre | nr | 0.30 | 4.50 | 17.16 | 3.25 | 24.91 |
| pubescens. Subsp. microphylla Superba, dwarf, pot 3 litre | nr | 0.20 | 3.00 | 7.16 | 1.52 | 11.68 |
| pubescens. Subsp. microphylla Superba, dwarf, pot 10 litre | nr | 0.30 | 4.50 | 17.16 | 3.25 | 24.91 |
| v. Charles Joly, dwarf, pot 3 litre | nr | 0.20 | 3.00 | 7.16 | 1.52 | 11.68 |
| v. Charles Joly, small, pot 10 litre | nr | 0.30 | 4.50 | 17.16 | 3.25 | 24.91 |
| v. Madame Lemoine, dwarf, pot 3 litre | nr | 0.20 | 3.00 | 7.16 | 1.52 | 11.68 |
| v. Madame Lemoine, small, pot 10 litre | nr | 0.30 | 4.50 | 17.16 | 3.25 | 24.91 |
| v. Michael Buchner, small, pot 10 litre | nr | 0.30 | 4.50 | 17.46 | 3.29 | 25.25 |

| | Unit | Labour | Hours £ | Mat'ls £ | O & P £ | Total £ |
|---|---|---|---|---|---|---|

**Specimen shrubs**

**The following descriptions refer to dimensions and sizes of shrubs as as set out below.**

Dwarf: 20 to 60cm high
Small: 60 to 120cm high
Medium: 120 to 200cm high
Large: over 200cm high

**Form planting hole in cultivated area, place shrub in hole, backfill, water and surround with peat**

| | Unit | Labour | Hours £ | Mat'ls £ | O & P £ | Total £ |
|---|---|---|---|---|---|---|
| Acer (Maple) | | | | | | |
| p. Fireglow, medium, pot 25 litre | nr | 0.60 | 9.00 | 74.83 | 12.57 | 96.40 |
| p. Fireglow, medium, pot 50 litre | nr | 1.00 | 15.00 | 124.72 | 20.96 | 160.68 |
| p. Fireglow, large, pot 130 litre | nr | 1.80 | 27.00 | 249.45 | 41.47 | 317.92 |
| Betula (Birch) | | | | | | |
| pendula, large, pot 55 litre | nr | 1.00 | 15.00 | 53.04 | 10.21 | 78.25 |
| Prunus (Cherry) | | | | | | |
| laurocerasus, medium, pot 15 litre | nr | 0.40 | 6.00 | 56.13 | 9.32 | 71.45 |
| laurocerasus, medium, pot 18 litre | nr | 0.40 | 6.00 | 74.83 | 12.12 | 92.95 |

| | Unit | Labour | Hours £ | Mat'ls £ | O & P £ | Total £ |
|---|---|---|---|---|---|---|
| laurocerasus, medium, pot 25 litre | nr | 0.60 | 9.00 | 111.70 | 18.11 | 138.81 |
| laurocerasus, medium, pot 55 litre | nr | 1.00 | 15.00 | 149.66 | 24.70 | 189.36 |
| lusitanica, medium, pot 20 litre | nr | 0.60 | 9.00 | 81.06 | 13.51 | 103.57 |
| lusitanica, medium, pot 25 litre | nr | 0.60 | 9.00 | 106.01 | 17.25 | 132.26 |

**CLIMBERS**

**The following descriptions refer to dimensions and sizes of shrubs as as set out below.**

Dwarf: 20 to 60cm high
Small: 60 to 120cm high
Medium: 120 to 200cm high
Large: over 200cm high

**Form planting hole in cultivated area, place climber in hole, backfill, water and surround with peat**

Clematis

| | Unit | Labour | Hours £ | Mat'ls £ | O & P £ | Total £ |
|---|---|---|---|---|---|---|
| Blue Bird, small, pot 2 litre | nr | 0.10 | 1.50 | 6.24 | 1.16 | 8.90 |
| x durandi, small, pot 2 litre | nr | 0.10 | 1.50 | 6.24 | 1.16 | 8.90 |
| Huldine, small, pot 2 litre | nr | 0.10 | 1.50 | 6.24 | 1.16 | 8.90 |
| Jackmanii, small, pot 2 litre | nr | 0.10 | 1.50 | 6.24 | 1.16 | 8.90 |

| | Unit | Labour | Hours £ | Mat'ls £ | O & P £ | Total £ |
|---|---|---|---|---|---|---|
| **Climber planting (cont'd)** | | | | | | |
| Laserstern, small, pot 2 litre | nr | 0.10 | 1.50 | 6.24 | 1.16 | 8.90 |
| Miss Bateman, small, pot 2 litre | nr | 0.10 | 1.50 | 6.24 | 1.16 | 8.90 |
| Nelly Moser, small, pot 2 litre | nr | 0.10 | 1.50 | 6.24 | 1.16 | 8.90 |
| Rouge Cardinal, small, pot 2 litre | nr | 0.10 | 1.50 | 6.24 | 1.16 | 8.90 |
| Sunset, small, pot 2 litre | nr | 0.10 | 1.50 | 6.24 | 1.16 | 8.90 |
| The President, small, | | | | 6.24 | | |
| pot 2 litre | nr | 0.10 | 1.50 | 6.24 | 1.16 | 8.90 |
| Vyvyan Pennell, small, pot 2 litre | nr | 0.10 | 1.50 | 6.24 | 1.16 | 8.90 |
| Wyevale, small, pot 2 litre | nr | 0.10 | 1.50 | 6.24 | 1.16 | 8.90 |
| Clematis Species | | | | | | |
| Alba Luxurians, small, pot 2 litre | nr | 0.10 | 1.50 | 6.24 | 1.16 | 8.90 |
| armandii, small, pot 2 litre | nr | 0.10 | 1.50 | 6.24 | 1.16 | 8.90 |
| a. Apple Blossom, small, pot 2 litre | nr | 0.10 | 1.50 | 6.24 | 1.16 | 8.90 |
| flamula, small, pot 2 litre | nr | 0.10 | 1.50 | 6.24 | 1.16 | 8.90 |
| montana, small, pot 2 litre | nr | 0.10 | 1.50 | 6.24 | 1.16 | 8.90 |
| m. Grandiflora, small, pot 2 litre | nr | 0.10 | 1.50 | 6.24 | 1.16 | 8.90 |
| m. var. rubens, small, pot 2 litre | nr | 0.10 | 1.50 | 6.24 | 1.16 | 8.90 |

| | Unit | Labour | Hours £ | Mat'ls £ | O & P £ | Total £ |
|---|---|---|---|---|---|---|
| montana var. sericea, small, pot 2 litre | nr | 0.10 | 1.50 | 6.24 | 1.16 | 8.90 |
| o. Orange Peel, small, pot 2 litre | nr | 0.10 | 1.50 | 6.24 | 1.16 | 8.90 |
| tangutica, small, pot 2 litre | nr | 0.10 | 1.50 | 6.24 | 1.16 | 8.90 |
| Venosa Violacea, small pot 2 litre | nr | 0.10 | 1.50 | 6.24 | 1.16 | 8.90 |
| vitalba, small, pot 2 litre | nr | 0.10 | 1.50 | 6.24 | 1.16 | 8.90 |
| viticella, small, pot 2 litre | nr | 0.10 | 1.50 | 6.24 | 1.16 | 8.90 |
| Hedera (Ivy) | | | | | | |
| canariensis, dwarf, pot 0.5 litre | nr | 0.10 | 1.50 | 1.51 | 0.45 | 3.46 |
| canariensis, small, pot 2 litre | nr | 0.10 | 1.50 | 4.66 | 0.92 | 7.08 |
| c. ravensholt, dwarf, pot 0.5 litre | nr | 0.10 | 1.50 | 1.51 | 0.45 | 3.46 |
| c. ravensholt, small, pot 2 litre | nr | 0.10 | 1.50 | 4.66 | 0.92 | 7.08 |
| c. dentata, dwarf, pot pot 0.5 litre | nr | 0.10 | 1.50 | 1.51 | 0.45 | 3.46 |
| c. dentata, small, pot 2 litre | nr | 0.10 | 1.50 | 4.66 | 0.92 | 7.08 |
| c. Sulphur Heart, dwarf, pot 0.5 litre | nr | 0.10 | 1.50 | 1.51 | 0.45 | 3.46 |
| c. Sulphur Heart, small, pot 2 litre | nr | 0.10 | 1.50 | 4.66 | 0.92 | 7.08 |
| helix, dwarf, pot 0.5 litre | nr | 0.10 | 1.50 | 1.51 | 0.45 | 3.46 |
| helix, small, pot 2 litre | nr | 0.10 | 1.50 | 4.66 | 0.92 | 7.08 |

| | Unit | Labour | Hours £ | Mat'ls £ | O & P £ | Total £ |
|---|---|---|---|---|---|---|
| **Climber planting (cont'd)** | | | | | | |
| h. buttercup, dwarf, pot 0.5 litre | nr | 0.10 | 1.50 | 1.51 | 0.45 | 3.46 |
| h. buttercup, small, pot 2 litre | nr | 0.10 | 1.50 | 4.66 | 0.92 | 7.08 |
| h. Goldchild, dwarf, pot 0.5 litre | nr | 0.10 | 1.50 | 1.51 | 0.45 | 3.46 |
| h. Goldchild, small, pot 2 litre | nr | 0.10 | 1.50 | 4.66 | 0.92 | 7.08 |
| c. Green Ripple, dwarf, pot 0.5 litre | nr | 0.10 | 1.50 | 1.51 | 0.45 | 3.46 |
| c. Green Ripple, small, pot 2 litre | nr | 0.10 | 1.50 | 4.66 | 0.92 | 7.08 |
| h. Ivalace, dwarf, pot 0.5 litre | nr | 0.10 | 1.50 | 1.51 | 0.45 | 3.46 |
| h. Ivalace, small, pot 2 litre | nr | 0.10 | 1.50 | 4.66 | 0.92 | 7.08 |
| h. Kolibri, dwarf, pot 0.5 litre | nr | 0.10 | 1.50 | 1.51 | 0.45 | 3.46 |
| h. Kolibri, small, pot 2 litre | nr | 0.10 | 1.50 | 4.66 | 0.92 | 7.08 |
| c. Light Fingers, dwarf, pot 0.5 litre | nr | 0.10 | 1.50 | 1.51 | 0.45 | 3.46 |
| c. Light Fingers, small, pot 2 litre | nr | 0.10 | 1.50 | 4.66 | 0.92 | 7.08 |
| c. Little Diamond, dwarf, pot 0.5 litre | nr | 0.10 | 1.50 | 1.51 | 0.45 | 3.46 |
| c. Little Diamond, small, pot 2 litre | nr | 0.10 | 1.50 | 4.66 | 0.92 | 7.08 |
| h. Luzii, dwarf, pot 0.5 litre | nr | 0.10 | 1.50 | 1.51 | 0.45 | 3.46 |
| h. Luzii, small, pot 2 litre | nr | 0.10 | 1.50 | 1.51 | 0.45 | 3.46 |

| | Unit | Labour | Hours £ | Mat'ls £ | O & P £ | Total £ |
|---|---|---|---|---|---|---|
| Hydrangea | | | | | | |
| anomala ssp. petiolaris, dwarf, pot 3 litre | nr | 0.10 | 1.50 | 6.24 | 1.16 | 8.90 |
| anomala ssp. petiolaris, small, pot 10 litre | nr | 0.20 | 3.00 | 13.71 | 2.51 | 19.22 |
| seemannii, small, pot pot 2 litre | nr | 0.10 | 1.50 | 1.51 | 0.45 | 3.46 |
| Jasminium (Jasmine) | | | | | | |
| humile Revolutum, dwarf, pot 2 litre | nr | 0.10 | 1.50 | 4.98 | 0.97 | 7.45 |
| nudiflorum, dwarf, pot pot 2 litre | nr | 0.20 | 3.00 | 4.98 | 1.20 | 9.18 |
| nudiflorum, small, pot pot 10 litre | nr | 0.20 | 3.00 | 13.71 | 2.51 | 19.22 |
| o. aureum, small, pot pot 2 litre | nr | 0.10 | 1.50 | 4.98 | 0.97 | 7.45 |
| o. Fiona Sunrise, small, pot 2 litre | nr | 0.10 | 1.50 | 6.24 | 1.16 | 8.90 |
| o. Inverleith, small, pot 2 litre | nr | 0.10 | 1.50 | 4.98 | 0.97 | 7.45 |
| x. stephanense, small, pot 2 litre | nr | 0.10 | 1.50 | 4.98 | 0.97 | 7.45 |
| Lonicera (Honeysuckle) | | | | | | |
| x. brownii, small, pot 2 litre | nr | 0.10 | 1.50 | 4.98 | 0.97 | 7.45 |
| x. heckrotii, small, pot 2 litre | nr | 0.10 | 1.50 | 4.98 | 0.97 | 7.45 |
| henryi, small, pot 2 litre | nr | 0.10 | 1.50 | 4.98 | 0.97 | 7.45 |
| x. italica, small, pot 2 litre | nr | 0.10 | 1.50 | 4.98 | 0.97 | 7.45 |

| | Unit | Labour | Hours £ | Mat'ls £ | O & P £ | Total £ |
|---|---|---|---|---|---|---|
| **Climber planting (cont'd)** | | | | | | |
| japonica Aureoreticulata, small, pot 2 litre | nr | 0.10 | 1.50 | 4.98 | 0.97 | 7.45 |
| j. Halliana, small, pot 2 litre | nr | 0.10 | 1.50 | 4.98 | 0.97 | 7.45 |
| periclymenum, small, pot 2 litre | nr | 0.10 | 1.50 | 4.98 | 0.97 | 7.45 |
| p. Belgica, small, pot 2 litre | nr | 0.10 | 1.50 | 4.98 | 0.97 | 7.45 |
| sempervirens, small, pot 2 litre | nr | 0.10 | 1.50 | 4.98 | 0.97 | 7.45 |
| Passiflora (Passion flower) | | | | | | |
| caerulea, small, pot 2 litre | nr | 0.10 | 1.50 | 4.98 | 0.97 | 7.45 |
| c. rubra, small, pot 2 litre | nr | 0.10 | 1.50 | 4.98 | 0.97 | 7.45 |
| Rosa (Rose) | | | | | | |
| banksiae Lutea, small, pot 4 litre | nr | 0.15 | 2.25 | 5.31 | 1.13 | 8.69 |
| Mermaid, small, pot 4 litre | nr | 0.15 | 2.25 | 5.31 | 1.13 | 8.69 |
| Wisteria | | | | | | |
| Burford, small, pot 3 litre | nr | 0.15 | 2.25 | 9.98 | 1.83 | 14.06 |
| Caroline, small, pot 3 litre | nr | 0.15 | 2.25 | 9.98 | 1.83 | 14.06 |
| Caroline, medium, pot 10 litre | nr | 0.20 | 3.00 | 18.71 | 3.26 | 24.97 |

| | Unit | Labour | Hours<br>£ | Mat'ls<br>£ | O & P<br>£ | Total<br>£ |
|---|---|---|---|---|---|---|
| floribunda Alba, small, pot 3 litre | nr | 0.15 | 2.25 | 9.98 | 1.83 | 14.06 |
| floribunda Alba, medium, pot 10 litre | nr | 0.20 | 3.00 | 18.71 | 3.26 | 24.97 |
| f. Domimo, small, pot 3 litre | nr | 0.15 | 2.25 | 9.98 | 1.83 | 14.06 |
| f. Domimo, medium, pot 10 litre | nr | 0.20 | 3.00 | 18.71 | 3.26 | 24.97 |
| f. Kuchi-beni, small, pot 3 litre | nr | 0.15 | 2.25 | 9.98 | 1.83 | 14.06 |
| f. Lawrence, small, pot 3 litre | nr | 0.15 | 2.25 | 9.98 | 1.83 | 14.06 |
| f. Multijuga, small, pot 3 litre | nr | 0.15 | 2.25 | 9.98 | 1.83 | 14.06 |
| f. Rosea, small, pot 3 litre | nr | 0.15 | 2.25 | 9.98 | 1.83 | 14.06 |
| f. Rosea, medium, pot 10 litre | nr | 0.20 | 3.00 | 18.71 | 3.26 | 24.97 |
| sinensis, small, pot 3 litre | nr | 0.15 | 2.25 | 9.98 | 1.83 | 14.06 |
| s. Alba, small, pot 3 litre | nr | 0.15 | 2.25 | 9.98 | 1.83 | 14.06 |
| s. Alba, medium, pot 10 litre | nr | 0.20 | 3.00 | 18.71 | 3.26 | 24.97 |
| s. Amethyst, small, pot 3 litre | nr | 0.15 | 2.25 | 9.98 | 1.83 | 14.06 |
| s. Amethyst, medium, pot 10 litre | nr | 0.20 | 3.00 | 18.71 | 3.26 | 24.97 |
| s. Blue Sapphire, small, pot 3 litre | nr | 0.15 | 2.25 | 9.96 | 1.83 | 14.04 |
| s. Prolific, small, pot 3 litre | nr | 0.15 | 2.25 | 9.96 | 1.83 | 14.04 |
| s. Prolific, medium, pot 10 litre | nr | 0.20 | 3.00 | 18.71 | 3.26 | 24.97 |

| | Unit | Labour | Hours £ | Mat'ls £ | O & P £ | Total £ |
|---|---|---|---|---|---|---|
| **HERBACEOUS PLANTS** | | | | | | |
| **Form planting hole in cultivated area, place plant in hole, backfill, water and surround with peat** | | | | | | |
| Anenome | | | | | | |
| hupehensis, 750mm high, pot 2 litre | nr | 0.15 | 2.25 | 3.74 | 0.90 | 6.89 |
| h. var. japonica, 1000mm high, pot 2 litre | nr | 0.15 | 2.25 | 3.74 | 0.90 | 6.89 |
| h. Splendens, 800mm high, pot 2 litre | nr | 0.15 | 2.25 | 3.74 | 0.90 | 6.89 |
| h. Elegans, 800mm high, pot 2 litre | nr | 0.15 | 2.25 | 3.74 | 0.90 | 6.89 |
| x h. Robustissima, 1200mm high, pot 2 litre | nr | 0.15 | 2.25 | 3.74 | 0.90 | 6.89 |
| x h. September, 750mm high, pot 2 litre | nr | 0.15 | 2.25 | 3.74 | 0.90 | 6.89 |
| x h. Whirlwind, 900mm high, pot 2 litre | nr | 0.15 | 2.25 | 3.74 | 0.90 | 6.89 |
| sylvestis Macrantha, 250mm high, pot 2 litre | nr | 0.15 | 2.25 | 3.74 | 0.90 | 6.89 |
| Campanula (Bell flower) | | | | | | |
| carpatica, 200mm high, pot 1 litre | nr | 0.10 | 1.50 | 2.50 | 0.60 | 4.60 |
| c. White Clips, 200mm high, pot 1 litre | nr | 0.10 | 1.50 | 2.50 | 0.60 | 4.60 |

| | Unit | Labour | Hours £ | Mat'ls £ | O & P £ | Total £ |
|---|---|---|---|---|---|---|
| cochlearifokia, 100mm high, pot 1 litre | nr | 0.10 | 1.50 | 1.25 | 0.41 | 3.16 |
| c. White Baby, 100mm high, pot 0.5 litre | nr | 0.10 | 1.50 | 1.25 | 0.41 | 3.16 |
| g. var.alba, 300mm high, pot 0.5 litre | nr | 0.10 | 1.50 | 1.90 | 0.51 | 3.91 |
| Kent Belle, 600mm high, pot 2 litre | nr | 0.15 | 2.25 | 3.42 | 0.85 | 6.52 |
| lactiflora Loddon Anna, 1200 high, pot 0.5 litre | nr | 0.10 | 1.50 | 1.25 | 0.41 | 3.16 |
| lactiflora Loddon Anna, 1500 high, pot 2 litre | nr | 0.15 | 2.25 | 2.50 | 0.71 | 5.46 |
| l. Violet, 1300mm high pot 0.5 litre | nr | 0.10 | 1.50 | 1.25 | 0.41 | 3.16 |
| l. Violet, 1300mm high pot 2 litre | nr | 0.15 | 2.25 | 3.42 | 0.85 | 6.52 |
| l. White Pouffe, 250mm high, pot 2 litre | nr | 0.15 | 2.25 | 2.50 | 0.71 | 5.46 |
| persificolia Alba, 900 high, pot 0.5 litre | nr | 0.15 | 2.25 | 1.25 | 0.53 | 4.03 |
| p. Caerulea Plena, 750mm high, pot 0.5 litre | nr | 0.15 | 2.25 | 1.25 | 0.53 | 4.03 |
| p. Telham Beauty, 900mm high, pot 2 litre | nr | 0.15 | 2.25 | 1.25 | 0.53 | 4.03 |
| p. Stella, 250mm high, pot 1 litre | nr | 0.15 | 2.25 | 2.50 | 0.71 | 5.46 |
| punctata, 300mm high, pot 2 litre | nr | 0.15 | 2.25 | 2.50 | 0.71 | 5.46 |
| takesimana, 600mm high, pot 2 litre | nr | 0.15 | 2.25 | 2.50 | 0.71 | 5.46 |
| l. Elizabeth, 600mm high, pot 2 litre | nr | 0.15 | 2.25 | 3.42 | 0.85 | 6.52 |

| | Unit | Labour | Hours £ | Mat'ls £ | O & P £ | Total £ |
|---|---|---|---|---|---|---|
| **Herbaceous plants planting (cont'd)** | | | | | | |
| Delphinium | | | | | | |
| Astolat Group, 1500mm high, pot 0.5 litre | nr | 0.10 | 1.50 | 1.25 | 0.41 | 3.16 |
| Astolat Group, 1500mm high, pot 2 litre | nr | 0.15 | 2.25 | 3.15 | 0.81 | 6.21 |
| Black Knight Group, 1800mm high, pot 0.5 litre | nr | 0.10 | 1.50 | 1.25 | 0.41 | 3.16 |
| Black Knight Group, 1800mm high, pot 2 litre | nr | 0.15 | 2.25 | 3.15 | 0.81 | 6.21 |
| Blue Bird Group, 1500mm high, pot 0.5 litre | nr | 0.10 | 1.50 | 1.25 | 0.41 | 3.16 |
| Blue Bird Group, 1500mm high, pot 2 litre | nr | 0.15 | 2.25 | 3.15 | 0.81 | 6.21 |
| Galahad Group, 1500mm high, pot 0.5 litre | nr | 0.10 | 1.50 | 1.25 | 0.41 | 3.16 |
| Galahad Group, 1500mm high, pot 2 litre | nr | 0.15 | 2.25 | 3.15 | 0.81 | 6.21 |
| grandiflorum, 400mm high, pot 0.5 litre | nr | 0.10 | 1.50 | 1.25 | 0.41 | 3.16 |
| grandiflorum, 400mm high, pot 2 litre | nr | 0.15 | 2.25 | 3.15 | 0.81 | 6.21 |
| King Arthur Group, 1500mm high, pot 0.5 litre | nr | 0.10 | 1.50 | 1.25 | 0.41 | 3.16 |
| King Arthur Group, 1500mm high, pot 2 litre | nr | 0.15 | 2.25 | 3.15 | 0.81 | 6.21 |

| | Unit | Labour | Hours £ | Mat'ls £ | O & P £ | Total £ |
|---|---|---|---|---|---|---|
| Summer Skies Group, 1500mm high, pot 0.5 litre | nr | 0.10 | 1.50 | 3.15 | 0.70 | 5.35 |
| Summer Skies Group, 1500mm high, pot 2 litre | nr | 0.15 | 2.25 | 1.25 | 0.53 | 4.03 |
| Geranium | | | | | | |
| Anne Thomson, 600mm high, pot 2 litre | nr | 0.15 | 2.25 | 3.42 | 0.85 | 6.52 |
| Bertie Crug, 200mm high, pot 2 litre | nr | 0.15 | 2.25 | 3.42 | 0.85 | 6.52 |
| x cantabrigiense, 250mm high, pot 2 litre | nr | 0.15 | 2.25 | 2.33 | 0.69 | 5.27 |
| cinerum Ballerina, 100mm high, pot 0.5 litre | nr | 0.10 | 1.50 | 1.95 | 0.52 | 3.97 |
| cinerum Ballerina, 100mm high, pot 2 litre | nr | 0.10 | 1.50 | 4.07 | 0.84 | 6.41 |
| c. var. subcaulescens, Splendens, 150mm high, pot 0.5 litre | nr | 0.10 | 1.50 | 1.95 | 0.52 | 3.97 |
| c. var. subcaulescens, Splendens, 150mm high, pot 2 litre | nr | 0.15 | 2.25 | 4.07 | 0.95 | 7.27 |
| c. Kashmir Purple, 300mm high, pot 2 litre | nr | 0.15 | 2.25 | 3.15 | 0.81 | 6.21 |
| c. Kashmir White, 300mm high, pot 2 litre | nr | 0.15 | 2.25 | 3.15 | 0.81 | 6.21 |
| Johnson's Blue, 150mm high, pot 0.5 litre | nr | 0.10 | 1.50 | 1.63 | 0.47 | 3.60 |
| Johnson's Blue, 150mm high, pot 2 litre | nr | 0.15 | 2.25 | 3.15 | 0.81 | 6.21 |

| | Unit | Labour | Hours £ | Mat'ls £ | O & P £ | Total £ |
|---|---|---|---|---|---|---|
| **Herbaceous plants planting (cont'd)** | | | | | | |
| macrorrhizum Album, 300mm high, pot 0.5 litre | nr | 0.10 | 1.50 | 1.84 | 0.50 | 3.84 |
| macrorrhizum Album, 300mm high, pot 2 litre | nr | 0.15 | 2.25 | 3.42 | 0.85 | 6.52 |
| m. Variegatum, 300mm high, pot 2 litre | nr | 0.15 | 2.25 | 3.42 | 0.85 | 6.52 |
| m. magnificum, 600mm high, pot 2 litre | nr | 0.15 | 2.25 | 2.33 | 0.69 | 5.27 |
| x. oxonianum, 450mm high, pot 2 litre | nr | 0.15 | 2.25 | 2.33 | 0.69 | 5.27 |
| x o. Wargrove Pink, 400mm high, pot 0.5 litre | nr | 0.10 | 1.50 | 1.25 | 0.41 | 3.16 |
| x o. Wargrove Pink, 400mm high, pot 2 litre | nr | 0.15 | 2.25 | 2.33 | 0.69 | 5.27 |
| Patricia, 750mm high, pot 2 litre | nr | 0.15 | 2.25 | 3.42 | 0.85 | 6.52 |
| phaeum, 800mm high, pot 2 litre | nr | 0.15 | 2.25 | 3.42 | 0.85 | 6.52 |
| p. Samobor, 400mm high, pot 2 litre | nr | 0.15 | 2.25 | 3.42 | 0.85 | 6.52 |
| pstilostemon, 1200mm high, pot 2 litre | nr | 0.15 | 2.25 | 3.42 | 0.85 | 6.52 |
| renardi, 250mm high, pot 2 litre | nr | 0.15 | 2.25 | 3.42 | 0.85 | 6.52 |
| x r. Russell Prichard, 200mm high, pot 0.5 litre | nr | 0.10 | 1.50 | 1.95 | 0.52 | 3.97 |

| | Unit | Labour | Hours £ | Mat'ls £ | O & P £ | Total £ |
|---|---|---|---|---|---|---|
| x r. Russell Prichard, 200mm high, pot 2 litre | nr | 0.15 | 2.25 | 4.07 | 0.95 | 7.27 |
| Salome, 450mm high, pot 2 litre | nr | 0.15 | 2.25 | 3.42 | 0.85 | 6.52 |
| s. Album, 300mm high, pot 2 litre | nr | 0.15 | 2.25 | 3.42 | 0.85 | 6.52 |
| s. Max Frei, 200mm high, pot 2 litre | nr | 0.15 | 2.25 | 3.42 | 0.85 | 6.52 |
| Spinners, 900mm high, pot 2 litre | nr | 0.15 | 2.25 | 3.42 | 0.85 | 6.52 |
| Sue Crug, 300mm high, pot 2 litre | nr | 0.15 | 2.25 | 3.42 | 0.85 | 6.52 |
| s. Mayflower, 300mm high, pot 2 litre | nr | 0.15 | 2.25 | 3.42 | 0.85 | 6.52 |
| Westray, 300mm high, pot 2 litre | nr | 0.15 | 2.25 | 3.42 | 0.85 | 6.52 |
| Gypsophila | | | | | | |
| paniculata Bristol Fairy, 1200mm high, pot 0.5 litre | nr | 0.10 | 1.50 | 1.57 | 0.46 | 3.53 |
| paniculata Bristol Fairy, 1200mm high, pot 2 litre | nr | 0.15 | 2.25 | 3.74 | 0.90 | 6.89 |
| p.Flamingo, 900mm high, pot 0.5 litre | nr | 0.10 | 1.50 | 1.57 | 0.46 | 3.53 |
| p.Flamingo, 900mm high, pot 2 litre | nr | 0.15 | 2.25 | 3.74 | 0.90 | 6.89 |
| repens Alba, 900mm high, pot 0.5 litre | nr | 0.10 | 1.50 | 1.25 | 0.41 | 3.16 |
| r. rosea, 900mm high, pot 0.5 litre | nr | 0.10 | 1.50 | 1.25 | 0.41 | 3.16 |

| | Unit | Labour | Hours £ | Mat'ls £ | O & P £ | Total £ |
|---|---|---|---|---|---|---|
| **Herbaceous plants planting (cont'd)** | | | | | | |
| Iris | | | | | | |
| Amber Queen, 200mm high, pot 2 litre | nr | 0.15 | 2.25 | 3.42 | 0.85 | 6.52 |
| Blue Denim, 250mm high, pot 2 litre | nr | 0.15 | 2.25 | 3.42 | 0.85 | 6.52 |
| Lilli-white, 300mm high, pot 2 litre | nr | 0.15 | 2.25 | 3.42 | 0.85 | 6.52 |
| Orange Caper, 250mm high, pot 2 litre | nr | 0.15 | 2.25 | 3.42 | 0.85 | 6.52 |
| Black Swan, 1000mm high, pot 2 litre | nr | 0.15 | 2.25 | 3.42 | 0.85 | 6.52 |
| Braithwaite, 900mm high, pot 2 litre | nr | 0.15 | 2.25 | 3.42 | 0.85 | 6.52 |
| Jane Phillips, 1200mm high, pot 2 litre | nr | 0.15 | 2.25 | 3.42 | 0.85 | 6.52 |
| Rajah, 900mm high, pot 2 litre | nr | 0.15 | 2.25 | 3.42 | 0.85 | 6.52 |
| Sable, 900mm high, pot 2 litre | nr | 0.15 | 2.25 | 3.42 | 0.85 | 6.52 |
| Stepping Out, 1000mm high, pot 2 litre | nr | 0.15 | 2.25 | 3.42 | 0.85 | 6.52 |
| White Knight, 1000mm high, pot 2 litre | nr | 0.15 | 2.25 | 3.42 | 0.85 | 6.52 |
| Lupinus (Lupin) | | | | | | |
| Chandelier, 1000mm high, pot 2 litre | nr | 0.15 | 2.25 | 3.15 | 0.81 | 6.21 |
| My Castle, 1000mm high, pot 2 litre | nr | 0.15 | 2.25 | 3.15 | 0.81 | 6.21 |
| Noble Maiden, 1000mm high, pot 2 litre | nr | 0.15 | 2.25 | 3.15 | 0.81 | 6.21 |

| | Unit | Labour | Hours £ | Mat'ls £ | O & P £ | Total £ |
|---|---|---|---|---|---|---|
| The Chateleine, 1000mm high, pot 2 litre | nr | 0.15 | 2.25 | 3.15 | 0.81 | 6.21 |
| The Governor, 1000mm high, pot 2 litre | nr | 0.15 | 2.25 | 3.15 | 0.81 | 6.21 |
| Paeonia (Peony) | | | | | | |
| lactiflora, 900mm high, pot 3 litre | nr | 0.15 | 2.25 | 5.31 | 1.13 | 8.69 |
| l. Duchesse de Nemours, 1000mm high, pot 3 litre | nr | 0.15 | 2.25 | 5.31 | 1.13 | 8.69 |
| l. Festiva Maxima, 1000mm high, pot 3 litre | nr | 0.15 | 2.25 | 5.31 | 1.13 | 8.69 |
| l. Karl Rosenfeld, 1000mm high, pot 3 litre | nr | 0.15 | 2.25 | 5.31 | 1.13 | 8.69 |
| l. Sarah Bernhardt, 1000mm high, pot 3 litre | nr | 0.15 | 2.25 | 5.31 | 1.13 | 8.69 |
| officinalis Alba Plena, 750mm high, pot 3 litre | nr | 0.15 | 2.25 | 5.31 | 1.13 | 8.69 |
| o. Rosea Plena, 750mm high, pot 3 litre | nr | 0.15 | 2.25 | 5.31 | 1.13 | 8.69 |
| o. Rubra Plena, 750mm high, pot 3 litre | nr | 0.15 | 2.25 | 5.31 | 1.13 | 8.69 |
| Phlox | | | | | | |
| douglasii Ice Mountain, 100mm high, pot 1 litre | nr | 0.10 | 1.50 | 2.49 | 0.60 | 4.59 |

| | Unit | Labour | Hours £ | Mat'ls £ | O & P £ | Total £ |
|---|---|---|---|---|---|---|
| **Herbaceous plants planting (cont'd)** | | | | | | |
| paniculata Blue Paradise, 800mm high, pot 2 litre | nr | 0.15 | 2.25 | 3.42 | 0.85 | 6.52 |
| p. Fujiyama, 800mm high, pot 2 litre | nr | 0.15 | 2.25 | 3.42 | 0.85 | 6.52 |
| p. Jules Sandeau, 800mm high, pot 2 litre | nr | 0.15 | 2.25 | 3.42 | 0.85 | 6.52 |
| p. Kirchenfuerst, 800mm high, pot 2 litre | nr | 0.15 | 2.25 | 3.42 | 0.85 | 6.52 |
| p. Little Boy, 700mm high, pot 2 litre | nr | 0.15 | 2.25 | 3.42 | 0.85 | 6.52 |
| p. Little Laura, 600mm high, pot 2 litre | nr | 0.15 | 2.25 | 3.42 | 0.85 | 6.52 |
| p. Little Princess, 500mm high, pot 2 litre | nr | 0.15 | 2.25 | 3.42 | 0.85 | 6.52 |
| p. Mia Ruys, 800mm high, pot 2 litre | nr | 0.15 | 2.25 | 3.42 | 0.85 | 6.52 |
| p. Norah Leigh, 600mm high, pot 2 litre | nr | 0.15 | 2.25 | 3.42 | 0.85 | 6.52 |
| p. Orange Perfection, 750mm, high, pot 2 litre | nr | 0.15 | 2.25 | 3.42 | 0.85 | 6.52 |
| p. Rijnstroom, 900mm high, pot 2 litre | nr | 0.15 | 2.25 | 3.42 | 0.85 | 6.52 |
| p. Starfire, 900mm high, pot 2 litre | nr | 0.15 | 2.25 | 3.42 | 0.85 | 6.52 |
| p. White Admiral, 900mm, high, pot 0.5 litre | nr | 0.10 | 1.50 | 1.25 | 0.41 | 3.16 |
| x procumbens Variegata, 200mm, high, pot 2 litre | nr | 0.15 | 2.25 | 3.47 | 0.86 | 6.58 |
| s. Emerald Cushion Blue, 200mm, high, pot 1 litre | nr | 0.10 | 1.50 | 2.49 | 0.60 | 4.59 |

| | Unit | Labour | Hours £ | Mat'ls £ | O & P £ | Total £ |
|---|---|---|---|---|---|---|
| Primula | | | | | | |
| x bulleesiana, 500mm high, pot 2 litre | nr | 0.15 | 2.25 | 2.82 | 0.76 | 5.83 |
| bulleyana, 600mm high, pot 2 litre | nr | 0.15 | 2.25 | 2.82 | 0.76 | 5.83 |
| d. Lilac, 250mm high, pot 0.5 litre | nr | 0.10 | 1.50 | 1.25 | 0.41 | 3.16 |
| d. Lilac, 250mm high, pot 2 litre | nr | 0.15 | 2.25 | 2.82 | 0.76 | 5.83 |
| d. Ruby, 250mm high, pot 0.5 litre | nr | 0.10 | 1.50 | 1.25 | 0.41 | 3.16 |
| d. Ruby, 250mm high, pot 2 litre | nr | 0.15 | 2.25 | 2.82 | 0.76 | 5.83 |
| florindae, 600mm high, pot 0.5 litre | nr | 0.10 | 1.50 | 1.25 | 0.41 | 3.16 |
| florindae, 600mm high, pot 2 litre | nr | 0.15 | 2.25 | 2.82 | 0.76 | 5.83 |
| j. Postford White, 600mm, high, pot 2 litre | nr | 0.15 | 2.25 | 2.82 | 0.76 | 5.83 |
| pulverenta, 600mm high, pot 2 litre | nr | 0.15 | 2.25 | 2.82 | 0.76 | 5.83 |
| rosae, 150mm high, pot 0.5 litre | nr | 0.10 | 1.50 | 1.25 | 0.41 | 3.16 |
| rosae, 150mm high, pot 2 litre | nr | 0.15 | 2.25 | 2.82 | 0.76 | 5.83 |
| veris, 200mm high, pot 0.5 litre | nr | 0.10 | 1.50 | 1.36 | 0.43 | 3.29 |
| vialli, 300mm high, pot 0.5 litre | nr | 0.10 | 1.50 | 1.57 | 0.46 | 3.53 |
| vulgaris acaulis, 300mm high, pot 0.5 litre | nr | 0.10 | 1.50 | 1.57 | 0.46 | 3.53 |
| vulgaris acaulis, 150mm high, pot 0.5 litre | nr | 0.15 | 2.25 | 1.57 | 0.57 | 4.39 |

| | Unit | Labour | Hours £ | Mat'ls £ | O & P £ | Total £ |
|---|---|---|---|---|---|---|
| **Herbaceous plants planting (cont'd)** | | | | | | |
| vulgaris acaulis, 150mm high, pot 2 litre | nr | 0.15 | 2.25 | 3.15 | 0.81 | 6.21 |
| Wanda, 150mm high, pot 0.5 litre | nr | 0.10 | 1.50 | 1.25 | 0.41 | 3.16 |
| Salvia | | | | | | |
| nemorosa Amethyst, 650mm high, pot 2 litre | nr | 0.15 | 2.25 | 3.15 | 0.81 | 6.21 |
| n. Ostfriesland, 450mm high, pot 0.5 litre | nr | 0.10 | 1.50 | 1.25 | 0.41 | 3.16 |
| n. Ostfriesland, 450mm high, pot 2 litre | nr | 0.15 | 2.25 | 3.15 | 0.81 | 6.21 |
| n. Rose Queen, 450mm high, pot 0.5 litre | nr | 0.10 | 1.50 | 1.25 | 0.41 | 3.16 |
| officinalis, 600mm high pot 1 litre | nr | 0.10 | 1.50 | 2.50 | 0.60 | 4.60 |
| o. Aurea, 300mm high pot 1 litre | nr | 0.10 | 1.50 | 2.50 | 0.60 | 4.60 |
| o. Aurea, 300mm high pot 2 litre | nr | 0.10 | 1.50 | 3.74 | 0.79 | 6.03 |
| o. Icterina, 600mm high, pot 1 litre | nr | 0.10 | 1.50 | 2.50 | 0.60 | 4.60 |
| o. Icterina, 600mm high, pot 2 litre | nr | 0.15 | 2.25 | 3.74 | 0.90 | 6.89 |
| o. Purpurascens, 600mm high, pot 1 litre | nr | 0.10 | 1.50 | 2.50 | 0.60 | 4.60 |
| o. Purpurascens, 600mm high, pot 2 litre | nr | 0.15 | 2.25 | 3.74 | 0.90 | 6.89 |
| o. Tricolor, 600mm high, pot 1 litre | nr | 0.10 | 1.50 | 4.26 | 0.86 | 6.62 |
| o. Tricolor, 600mm | | | | | | |

| | Unit | Labour | Hours £ | Mat'ls £ | O & P £ | Total £ |
|---|---|---|---|---|---|---|
| high, pot 2 litre | nr | 0.15 | 2.25 | 3.74 | 0.90 | 6.89 |
| patens, 450mm high, pot 2 litre | nr | 0.15 | 2.25 | 3.74 | 0.90 | 6.89 |
| p. Cambridge Blue, 600mm high, pot 2 litre | nr | 0.15 | 2.25 | 3.74 | 0.90 | 6.89 |
| verticillata Alba, 600mm high, pot 2 litre | nr | 0.15 | 2.25 | 3.15 | 0.81 | 6.21 |
| v. Purple Rain, 450mm high, pot 2 litre | nr | 0.15 | 2.25 | 3.15 | 0.81 | 6.21 |
| Veronica | | | | | | |
| gentianoides, 300mm high, pot 1 litre | nr | 0.10 | 1.50 | 2.50 | 0.60 | 4.60 |
| peduncularis, 150mm high, pot 1 litre | nr | 0.10 | 1.50 | 2.50 | 0.60 | 4.60 |
| prostrata, 100mm high, pot 1 litre | nr | 0.10 | 1.50 | 2.50 | 0.60 | 4.60 |
| p. Trehane, 150mm high, pot 1 litre | nr | 0.10 | 1.50 | 2.50 | 0.60 | 4.60 |
| spicata, 750mm high, pot 2 litre | nr | 0.15 | 2.25 | 3.74 | 0.90 | 6.89 |
| s. Alba, 400mm high, pot 0.5 litre | nr | 0.10 | 1.50 | 1.25 | 0.41 | 3.16 |
| Viola (Violet) | | | | | | |
| Ardross Gem, 150mm high, pot 0.5 litre | nr | 0.10 | 1.50 | 1.25 | 0.41 | 3.16 |
| Ardross Gem, 150mm high, pot 1 litre | nr | 0.10 | 1.50 | 2.50 | 0.60 | 4.60 |
| Belmont Blue, 150mm high, pot 0.5 litre | nr | 0.10 | 1.50 | 1.25 | 0.41 | 3.16 |
| Belmont Blue, 150mm high, pot 1 litre | nr | 0.10 | 1.50 | 2.50 | 0.60 | 4.60 |

| | Unit | Labour | Hours £ | Mat'ls £ | O & P £ | Total £ |
|---|---|---|---|---|---|---|
| **Herbaceous plants planting (cont'd)** | | | | | | |
| Buttercup, 150mm high, pot 0.5 litre | nr | 0.10 | 1.50 | 1.25 | 0.41 | 3.16 |
| Buttercup, 150mm high, pot 1 litre | nr | 0.10 | 1.50 | 2.50 | 0.60 | 4.60 |
| Dawn, 150mm high, pot 0.5 litre | nr | 0.10 | 1.50 | 1.25 | 0.41 | 3.16 |
| Dawn, 150mm high, pot 1 litre | nr | 0.10 | 1.50 | 2.50 | 0.60 | 4.60 |
| Etain, 150mm high, pot 0.5 litre | nr | 0.10 | 1.50 | 1.25 | 0.41 | 3.16 |
| Etain, 150mm high, pot 1 litre | nr | 0.10 | 1.50 | 2.50 | 0.60 | 4.60 |
| Jackanapes, 150mm high, pot 0.5 litre | nr | 0.10 | 1.50 | 1.25 | 0.41 | 3.16 |
| Jackanapes, 150mm high, pot 1 litre | nr | 0.10 | 1.50 | 2.50 | 0.60 | 4.60 |
| Janet, 150mm high, pot 0.5 litre | nr | 0.10 | 1.50 | 1.25 | 0.41 | 3.16 |
| Janet, 150mm high, pot 1 litre | nr | 0.10 | 1.50 | 2.50 | 0.60 | 4.60 |
| Labradorica, 100mm high, pot 0.5 litre | nr | 0.10 | 1.50 | 1.25 | 0.41 | 3.16 |
| Maggie Matt, 150mm high, pot 0.5 litre | nr | 0.10 | 1.50 | 1.25 | 0.41 | 3.16 |
| Maggie Matt, 150mm high, pot 1 litre | nr | 0.10 | 1.50 | 2.50 | 0.60 | 4.60 |
| Mrs Lancaster, 150mm high, pot 0.5 litre | nr | 0.10 | 1.50 | 1.25 | 0.41 | 3.16 |
| Mrs Lancaster, 150mm high, pot 1 litre | nr | 0.10 | 1.50 | 2.50 | 0.60 | 4.60 |
| odorata, 100mm high, pot 0.5 litre | nr | 0.10 | 1.50 | 1.25 | 0.41 | 3.16 |

| | Unit | Labour | Hours £ | Mat'ls £ | O & P £ | Total £ |
|---|---|---|---|---|---|---|
| Rebecca, 150mm high, pot 0.5 litre | nr | 0.10 | 1.50 | 2.50 | 0.60 | 4.60 |
| Rebecca, 150mm high, pot 1 litre | nr | 0.10 | 1.50 | 1.25 | 0.41 | 3.16 |
| Roscastle Black, 150mm high, pot 0.5 litre | nr | 0.10 | 1.50 | 2.50 | 0.60 | 4.60 |
| Roscastle Black, 150mm high, pot 1 litre | nr | 0.10 | 1.50 | 1.25 | 0.41 | 3.16 |
| scoria Freckles, 150mm high, pot 1 litre | nr | 0.10 | 1.50 | 2.50 | 0.60 | 4.60 |
| Zoe, 150mm high, pot 1 litre | nr | 0.10 | 1.50 | 2.50 | 0.60 | 4.60 |

## HEDGING

**The following descriptions refer to dimensions and sizes of hedge plants as set out below.**

Dwarf: 20 to 60cm high
Small: 60 to 120cm high
Medium: 120 to 200cm high
Large: over 200cm high

**Form planting hole in cultivated area, place hedging plant in hole, backfill, water and surround with peat**

Acer (Maple)

| | Unit | Labour | Hours £ | Mat'ls £ | O & P £ | Total £ |
|---|---|---|---|---|---|---|
| campestre, small, pot 1 litre | nr | 0.10 | 1.50 | 0.60 | 0.32 | 2.42 |
| campestre, small, pot 1 litre | nr | 0.10 | 1.50 | 0.65 | 0.32 | 2.47 |

| | Unit | Labour | Hours £ | Mat'ls £ | O & P £ | Total £ |
|---|---|---|---|---|---|---|
| **Hedging (cont'd)** | | | | | | |
| campestre, medium, pot 1 litre | nr | 0.10 | 1.50 | 0.76 | 0.34 | 2.60 |
| campestre, medium, pot 2 litre | nr | 0.15 | 2.25 | 1.25 | 0.53 | 4.03 |
| platanoides, small, pot 1 litre | nr | 0.10 | 1.50 | 0.60 | 0.32 | 2.42 |
| platanoides, small, pot 1 litre | nr | 0.10 | 1.50 | 0.65 | 0.32 | 2.47 |
| platanoised, medium, pot 1 litre | nr | 0.10 | 1.50 | 0.76 | 0.34 | 2.60 |
| platanoides, medium, pot 2 litre | nr | 0.15 | 2.25 | 1.25 | 0.53 | 4.03 |
| platanoides, small, pot 1 litre | nr | 0.10 | 1.50 | 0.60 | 0.32 | 2.42 |
| platanoides, small, pot 1 litre | nr | 0.10 | 1.50 | 0.65 | 0.32 | 2.47 |
| platanoised, medium, pot 1 litre | nr | 0.10 | 1.50 | 0.76 | 0.34 | 2.60 |
| platanoides, medium, pot 2 litre | nr | 0.15 | 2.25 | 1.25 | 0.53 | 4.03 |
| pseudoplatanus, small, pot 1 litre | nr | 0.10 | 1.50 | 0.60 | 0.32 | 2.42 |
| pseudoplatanus, small, pot 1 litre | nr | 0.10 | 1.50 | 0.65 | 0.32 | 2.47 |
| pseudoplatanus, medium, pot 1 litre | nr | 0.15 | 2.25 | 0.76 | 0.45 | 3.46 |
| pseudoplatanus, medium, pot 2 litre | nr | 0.15 | 2.25 | 1.25 | 0.53 | 4.03 |
| Alnus (Alder) | | | | | | |
| cordata, small, pot 1 litre | nr | 0.10 | 1.50 | 0.60 | 0.32 | 2.42 |

| | Unit | Labour | Hours £ | Mat'ls £ | O & P £ | Total £ |
|---|---|---|---|---|---|---|
| cordata, small, pot 1 litre | nr | 0.10 | 1.50 | 0.65 | 0.32 | 2.47 |
| codarta, medium, pot 1 litre | nr | 0.10 | 1.50 | 0.76 | 0.34 | 2.60 |
| cordata, medium, pot 2 litre | nr | 0.15 | 2.25 | 1.25 | 0.53 | 4.03 |
| glutinosa, small, pot 1 litre | nr | 0.10 | 1.50 | 0.60 | 0.32 | 2.42 |
| glutinosa, small, pot 1 litre | nr | 0.10 | 1.50 | 0.65 | 0.32 | 2.47 |
| glutinosa, medium, pot 1 litre | nr | 0.10 | 1.50 | 0.76 | 0.34 | 2.60 |
| glutinosa, medium, pot 2 litre | nr | 0.15 | 2.25 | 1.25 | 0.53 | 4.03 |
| incana, small, pot 1 litre | nr | 0.10 | 1.50 | 0.60 | 0.32 | 2.42 |
| incana, small, pot 1 litre | nr | 0.10 | 1.50 | 0.65 | 0.32 | 2.47 |
| incana, medium, pot 1 litre | nr | 0.10 | 1.50 | 0.76 | 0.34 | 2.60 |
| incana, medium, pot 2 litre | nr | 0.15 | 2.25 | 1.25 | 0.53 | 4.03 |
| Betula (Birch) | | | | | | |
| pendula, small, pot 1 litre | nr | 0.10 | 1.50 | 0.60 | 0.32 | 2.42 |
| pendula, small, pot 1 litre | nr | 0.10 | 1.50 | 0.65 | 0.32 | 2.47 |
| pendula, medium, pot 1 litre | nr | 0.10 | 1.50 | 0.76 | 0.34 | 2.60 |
| pendula, medium, pot 2 litre | nr | 0.15 | 2.25 | 1.25 | 0.53 | 4.03 |

| | Unit | Labour | Hours £ | Mat'ls £ | O & P £ | Total £ |
|---|---|---|---|---|---|---|
| **Hedging (cont'd)** | | | | | | |
| pubescens, small, pot 1 litre | nr | 0.10 | 1.50 | 0.60 | 0.32 | 2.42 |
| pubescens, small, pot 1 litre | nr | 0.10 | 1.50 | 0.65 | 0.32 | 2.47 |
| pubescens, medium, pot 1 litre | nr | 0.10 | 1.50 | 0.76 | 0.34 | 2.60 |
| Chamaecyparis (Cypress) | | | | | | |
| lawsoniana, small, pot 3 litre | nr | 0.20 | 3.00 | 5.63 | 1.29 | 9.92 |
| lawsoniama, small, pot 10 litre | nr | 0.30 | 4.50 | 9.98 | 2.17 | 16.65 |
| Crataegus (Hawthorn) | | | | | | |
| laevigata, small, pot 1 litre | nr | 0.10 | 1.50 | 0.60 | 0.32 | 2.42 |
| laevigata, small, pot 1 litre | nr | 0.10 | 1.50 | 0.65 | 0.32 | 2.47 |
| laevigata, medium, pot 1 litre | nr | 0.10 | 1.50 | 0.76 | 0.34 | 2.60 |
| laevigata, medium, pot 2 litre | nr | 0.15 | 2.25 | 1.25 | 0.53 | 4.03 |
| Cupressocyparis | | | | | | |
| leylandii, small, pot 3 litre | nr | 0.20 | 3.00 | 3.74 | 1.01 | 7.75 |
| leylandii, medium, pot 10 litre | nr | 0.30 | 4.50 | 9.98 | 2.17 | 16.65 |
| l. Castlewellan, small, pot 3 litre | nr | 0.20 | 3.00 | 3.74 | 1.01 | 7.75 |

| | Unit | Labour | Hours £ | Mat'ls £ | O & P £ | Total £ |
|---|---|---|---|---|---|---|
| I. Castlewellan, small, pot 3 litre | nr | 0.20 | 3.00 | 3.74 | 1.01 | 7.75 |
| Fagus (Beech) | | | | | | |
| sylvatica, dwarf, pot 3 litre | nr | 0.20 | 3.00 | 2.50 | 0.83 | 6.33 |
| sylvatica, dwarf, pot 10 litre | nr | 0.30 | 4.50 | 12.47 | 2.55 | 19.52 |
| Fagus | | | | | | |
| pendula, small, pot 1 litre | nr | 0.10 | 1.50 | 0.76 | 0.34 | 2.60 |
| pendula, small, pot 1 litre | nr | 0.10 | 1.50 | 0.92 | 0.36 | 2.78 |
| pendula, medium, pot 2 litre | nr | 0.15 | 2.25 | 1.25 | 0.53 | 4.03 |
| pendula, medium, pot 2 litre | nr | 0.15 | 2.25 | 1.63 | 0.58 | 4.46 |
| Atropurpurea, small, pot 1 litre | nr | 0.10 | 1.50 | 2.00 | 0.53 | 4.03 |
| Atropurpurea, small, pot 1 litre | nr | 0.10 | 1.50 | 2.60 | 0.62 | 4.72 |
| Atropurpurea, medium, pot 2 litre | nr | 0.15 | 2.25 | 2.82 | 0.76 | 5.83 |
| Atropurpurea, medium, pot 2 litre | nr | 0.15 | 2.25 | 4.07 | 0.95 | 7.27 |
| Ligustrum (Privet) | | | | | | |
| ovalifolium, dwarf, pot 1 litre | nr | 0.10 | 1.50 | 0.55 | 0.31 | 2.36 |
| ovalifolium, small, pot 1 litre | nr | 0.10 | 1.50 | 0.60 | 0.32 | 2.42 |

| | Unit | Labour | Hours £ | Mat'ls £ | O & P £ | Total £ |
|---|---|---|---|---|---|---|
| **Hedging (cont'd)** | | | | | | |
| ovalifolium, small, pot 1 litre | nr | 0.10 | 1.50 | 0.76 | 0.34 | 2.60 |
| ovalifolium, small, pot 3 litre | nr | 0.20 | 3.00 | 2.50 | 0.83 | 6.33 |
| vulgare, dwarf, pot 1 litre | nr | 0.10 | 1.50 | 0.55 | 0.31 | 2.36 |
| vulgare, small, pot 1 litre | nr | 0.10 | 1.50 | 0.60 | 0.32 | 2.42 |
| vulgare, small, pot 1 litre | nr | 0.10 | 1.50 | 0.76 | 0.34 | 2.60 |
| vulgare, small, pot 3 litre | nr | 0.20 | 3.00 | 2.50 | 0.83 | 6.33 |
| Prunus (Cherry) | | | | | | |
| avium, small, pot 1 litre | nr | 0.10 | 1.50 | 0.60 | 0.32 | 2.42 |
| avium, small, pot 1 litre | nr | 0.10 | 1.50 | 0.76 | 0.34 | 2.60 |
| avium, medium, pot 1 litre | nr | 0.10 | 1.50 | 1.25 | 0.41 | 3.16 |
| cerasifera, small, pot 1 litre | nr | 0.10 | 1.50 | 0.55 | 0.31 | 2.36 |
| cerasifera, small, pot 1 litre | nr | 0.10 | 1.50 | 0.60 | 0.32 | 2.42 |
| cerasifera, medium, pot 1 litre | nr | 0.10 | 1.50 | 0.76 | 0.34 | 2.60 |
| laurocerasus, small, pot 3 litre | nr | 0.20 | 3.00 | 2.50 | 0.83 | 6.33 |
| laurocerasus, small, pot 3 litre | nr | 0.20 | 3.00 | 3.75 | 1.01 | 7.76 |
| laurocerasus, small, pot 10 litre | nr | 0.30 | 4.50 | 13.76 | 2.74 | 21.00 |

| | Unit | Labour | Hours £ | Mat'ls £ | O & P £ | Total £ |
|---|---|---|---|---|---|---|
| lusitanica, dwarf, pot 3 litre | nr | 0.20 | 3.00 | 3.75 | 1.01 | 7.76 |
| lusitanica, dwarf, pot 3 litre | nr | 0.20 | 3.00 | 4.35 | 1.10 | 8.45 |
| lusitanica, small, pot 10 litre | nr | 0.30 | 4.50 | 13.76 | 2.74 | 21.00 |
| padus, small, pot 1 litre | nr | 0.10 | 1.50 | 0.60 | 0.32 | 2.42 |
| padus, small, pot 1 litre | nr | 0.10 | 1.50 | 0.76 | 0.34 | 2.60 |
| padus, medium, pot 2 litre | nr | 0.15 | 2.25 | 1.25 | 0.53 | 4.03 |
| spinosa, dwarf, pot 1 litre | nr | 0.10 | 1.50 | 0.60 | 0.32 | 2.42 |
| spinosa, small, pot 1 litre | nr | 0.10 | 1.50 | 0.70 | 0.33 | 2.53 |
| spinosa, small, pot 1 litre | nr | 0.10 | 1.50 | 1.03 | 0.38 | 2.91 |
| Quercus (Oak) | | | | | | |
| cerris, dwarf, pot 1 litre | nr | 0.10 | 1.50 | 0.70 | 0.33 | 2.53 |
| cerris, small, pot 1 litre | nr | 0.10 | 1.50 | 0.92 | 0.36 | 2.78 |
| ilex, dwarf, pot 3 litre | nr | 0.20 | 3.00 | 3.75 | 1.01 | 7.76 |
| ilex, small, pot 3 litre | nr | 0.20 | 3.00 | 5.00 | 1.20 | 9.20 |
| ilex, medium, pot 10 litre | nr | 0.30 | 4.50 | 17.51 | 3.30 | 25.31 |
| palustris, dwarf, pot 1 litre | nr | 0.10 | 1.50 | 0.76 | 0.34 | 2.60 |
| palustris, small, pot 1 litre | nr | 0.10 | 1.50 | 0.92 | 0.36 | 2.78 |
| robur, dwarf, pot 1 litre | nr | 0.10 | 1.50 | 0.70 | 0.33 | 2.53 |
| robur, small, pot 2 litre | nr | 0.15 | 2.25 | 0.92 | 0.48 | 3.65 |

| | Unit | Labour | Hours £ | Mat'ls £ | O & P £ | Total £ |
|---|---|---|---|---|---|---|
| **Hedging (cont'd)** | | | | | | |
| robur, small, pot 1 litre | nr | 0.10 | 1.50 | 1.03 | 0.38 | 2.91 |
| robur, medium, pot 2 litre | nr | 0.15 | 2.25 | 1.63 | 0.58 | 4.46 |
| rubra, dwarf, pot 1 litre | nr | 0.10 | 1.50 | 0.70 | 0.33 | 2.53 |
| rubra, small, pot 2 litre | nr | 0.15 | 2.25 | 0.92 | 0.48 | 3.65 |
| rubra, small, pot 1 litre | nr | 0.10 | 1.50 | 1.03 | 0.38 | 2.91 |
| rubra, medium, pot 2 litre | nr | 0.15 | 2.25 | 1.63 | 0.58 | 4.46 |
| Salix (Willow) | | | | | | |
| alba, small, pot 1litre | nr | 0.10 | 1.50 | 0.60 | 0.32 | 2.42 |
| alba, small, pot 1 litre | nr | 0.10 | 1.50 | 0.76 | 0.34 | 2.60 |
| alba, medium, pot 1 litre | nr | 0.15 | 2.25 | 1.25 | 0.53 | 4.03 |
| a. Liempde, small, pot 1 litre | nr | 0.10 | 1.50 | 0.60 | 0.32 | 2.42 |
| a. Liempde, small, pot 1 litre | nr | 0.10 | 1.50 | 0.76 | 0.34 | 2.60 |
| alba var.caerulea, small, pot 1 litre | nr | 0.10 | 1.50 | 0.60 | 0.32 | 2.42 |
| alba var.caerulea, small, pot 1 litre | nr | 0.10 | 1.50 | 0.76 | 0.34 | 2.60 |
| a. subsp. vitellina, small, pot 1 litre | nr | 0.10 | 1.50 | 0.60 | 0.32 | 2.42 |
| a. subsp. vitellina, small, pot 1 litre | nr | 0.10 | 1.50 | 0.76 | 0.34 | 2.60 |
| a. subsp. vitellina, small, pot 3 litre | nr | 0.20 | 3.00 | 2.50 | 0.83 | 6.33 |
| caprea, small, pot 1litre | nr | 0.10 | 1.50 | 0.60 | 0.32 | 2.42 |
| caprea, small, pot 1 litre | nr | 0.10 | 1.50 | 0.76 | 0.34 | 2.60 |

| | Unit | Labour | Hours | Mat'ls | O & P | Total |
|---|---|---|---|---|---|---|
| caprea, medium, pot 1 litre | nr | 0.15 | 2.25 | 1.25 | 0.53 | 4.03 |
| cinerea, small, pot 1litre | nr | 0.10 | 1.50 | 0.60 | 0.32 | 2.42 |
| cinerea, small, pot 1 litre | nr | 0.10 | 1.50 | 0.76 | 0.34 | 2.60 |
| cinerea, small, pot 3 litre | nr | 0.20 | 3.00 | 2.50 | 0.83 | 6.33 |
| daphnoides, small, pot 1 litre | nr | 0.10 | 1.50 | 0.60 | 0.32 | 2.42 |
| daphnoides, small, pot 1 litre | nr | 0.10 | 1.50 | 0.76 | 0.34 | 2.60 |
| daphnoides, small, pot 3 litre | nr | 0.10 | 1.50 | 2.50 | 0.60 | 4.60 |
| elaeangnos, small, pot 1 litre | nr | 0.10 | 1.50 | 0.60 | 0.32 | 2.42 |
| elaeangnos, small, pot 3 litre | nr | 0.20 | 3.00 | 2.50 | 0.83 | 6.33 |
| viminalis, small, pot 1 litre | nr | 0.10 | 1.50 | 0.60 | 0.32 | 2.42 |
| viminalis, small, pot 1 litre | nr | 0.10 | 1.50 | 0.76 | 0.34 | 2.60 |
| viminalis, small, pot 3 litre | nr | 0.10 | 1.50 | 2.50 | 0.60 | 4.60 |
| Sorbus (Whitebeam) | | | | | | |
| aria, small, pot 1litre | nr | 0.10 | 1.50 | 0.76 | 0.34 | 2.60 |
| aria, small, pot 1litre | nr | 0.10 | 1.50 | 0.98 | 0.37 | 2.85 |
| aria, small, pot 1litre | nr | 0.10 | 1.50 | 1.20 | 0.41 | 3.11 |
| aria, medium, pot 1 litre | nr | 0.15 | 2.25 | 1.63 | 0.58 | 4.46 |
| acuparia, small, pot 1 litre | nr | 0.10 | 1.50 | 0.54 | 0.31 | 2.35 |
| acuparia, small, pot 1 litre | nr | 0.10 | 1.50 | 0.65 | 0.32 | 2.47 |
| acuparia, small, pot 1 litre | nr | 0.10 | 1.50 | 0.76 | 0.34 | 2.60 |

| | Unit | Labour | Hours £ | Mat'ls £ | O & P £ | Total £ |
|---|---|---|---|---|---|---|
| **Hedging (cont'd)** | | | | | | |
| acuparia, medium, pot 2 litre | nr | 0.15 | 2.25 | 1.25 | 0.53 | 4.03 |
| intermedia, small, pot 1 litre | nr | 0.10 | 1.50 | 0.76 | 0.34 | 2.60 |
| intermedia, small, pot 1 litre | nr | 0.10 | 1.50 | 0.98 | 0.37 | 2.85 |
| intermedia, small, pot 1 litre | nr | 0.10 | 1.50 | 1.20 | 0.41 | 3.11 |
| intermedia, medium, pot 2 litre | nr | 0.15 | 2.25 | 1.63 | 0.58 | 4.46 |

**BEDDING PLANTS**

The following items cover the work involved in the creation of a formal bedding display.The cost of supplying and bedding the plants has not been included because of the wide variation in density, design patern and quality of the plants involved.

| | Unit | Labour | Hours £ | Mat'ls £ | O & P £ | Total £ |
|---|---|---|---|---|---|---|
| Protect edge of existing lawn with polythene sheeting (6 uses) | m | 0.03 | 0.45 | 0.06 | 0.08 | 0.59 |
| Remove existing plants from bed | m2 | 1.40 | 21.00 | - | 3.15 | 24.15 |

| | Unit | Labour | Hours £ | Mat'ls £ | O & P £ | Total £ |
|---|---|---|---|---|---|---|
| Fork over to a depth of 300mm, remove weeds and rake over | m2 | 2.00 | 30.00 | - | 4.50 | 34.50 |
| Cut edge of existing lawn with shears | m | 0.03 | 0.45 | - | 0.07 | 0.52 |
| Dig trench 300mm deep and fork in well-rotted manure 100mm deep to bottom of trench | m | 0.12 | 1.80 | 0.16 | 0.29 | 2.25 |
| **MAINTENANCE WORK BY HAND** | | | | | | |
| **Grassed areas** | | | | | | |
| Pick up litter and remove | 100m2 | 0.15 | 2.25 | 1.63 | 0.58 | 4.46 |
| Rake up loose grass and place in heaps for disposal | 100m2 | 0.35 | 5.25 | 1.63 | 1.03 | 7.91 |
| Trim edges of grassed areas with edging tool | 100m2 | 0.50 | 7.50 | 1.63 | 1.37 | 10.50 |
| Aerate by forking | 100m2 | 1.50 | 22.50 | 1.63 | 3.62 | 27.75 |
| Sweep up corings and place in heaps for disposal | 100m2 | 1.50 | 22.50 | 1.63 | 3.62 | 27.75 |
| Apply weedkiller (175 kg per ha) | 100m2 | 0.50 | 7.50 | 0.33 | 1.17 | 9.00 |

| | Unit | Labour | Hours £ | Mat'ls £ | O & P £ | Total £ |
|---|---|---|---|---|---|---|
| Apply fertiliser (35g per) ha) | 100m2 | 0.50 | 7.50 | 0.33 | 1.17 | 9.00 |
| **Planted areas** | | | | | | |
| Remove stones and debris and place in heaps for disposal | 100m2 | 1.00 | 15.00 | 0.33 | 2.30 | 17.63 |
| Weed and hoe and place weeds in heaps for disposal | 100m2 | 1.00 | 15.00 | 0.33 | 2.30 | 17.63 |
| Fine-grade bark mulch, thickness | | | | | | |
| 50mm | 100m2 | 1.50 | 22.50 | 141.38 | 24.58 | 188.46 |
| 75mm | 100m2 | 1.75 | 26.25 | 212.07 | 35.75 | 274.07 |
| 100mm | 100m2 | 2.00 | 30.00 | 282.76 | 46.91 | 359.67 |
| Medium-grade bark mulch, thickness | | | | | | |
| 50mm | 100m2 | 1.50 | 22.50 | 204.45 | 34.04 | 260.99 |
| 75mm | 100m2 | 1.75 | 26.25 | 306.68 | 49.94 | 382.87 |
| 100mm | 100m2 | 2.00 | 30.00 | 398.04 | 64.21 | 492.25 |
| Coarse-grade bark mulch, thickness | | | | | | |
| 50mm | 100m2 | 1.50 | 22.50 | 243.61 | 39.92 | 306.03 |
| 75mm | 100m2 | 1.75 | 26.25 | 365.41 | 58.75 | 450.41 |
| 100mm | 100m2 | 2.00 | 30.00 | 487.21 | 77.58 | 594.79 |

| | Unit | Plant £ | Mat'ls £ | O & P £ | Total £ |
|---|---|---|---|---|---|
| **MAINTENANCE WORK BY MACHINE** | | | | | |
| Where applicable the plant column includes the cost of the operator. | | | | | |
| **Grassed areas** | | | | | |
| Sweep up leaves with motorised vacuum cleaner | 100m2 | 0.84 | - | 0.13 | 0.97 |
| Cut grass to specified height with | | | | | |
| petrol-powered mower | 100m2 | 2.40 | - | 0.36 | 2.76 |
| multi-unit mower | 100m2 | 0.48 | - | 0.07 | 0.55 |
| ride-on rotary mower | 100m2 | 0.67 | - | 0.10 | 0.77 |
| Aerate grassed surfaces with | | | | | |
| slitter aerator | 100m2 | 5.56 | - | 0.83 | 6.39 |
| tractor-drawn aeratot | 100m2 | 3.22 | - | 0.48 | 3.70 |
| Harrow grassed surfaces with | | | | | |
| chain harrow | 100m2 | 0.98 | - | 0.15 | 1.13 |
| drag mat | 100m2 | 1.10 | - | 0.17 | 1.27 |

| | Unit | Labour | Hours £ | Mat'ls £ | O & P £ | Total £ |
|---|---|---|---|---|---|---|
| **SUNDRIES** | | | | | | |
| Treated softwood tree stakes driven into the ground | | | | | | |
| 50mm diameter | | | | | | |
| 1.7m long | nr | 0.10 | 1.50 | 2.04 | 0.53 | 4.07 |
| 2.0m long | nr | 0.12 | 1.80 | 2.39 | 0.63 | 4.82 |
| 2.4m long | nr | 0.14 | 2.10 | 2.88 | 0.75 | 5.73 |
| 100mm diameter | | | | | | |
| 1.7m long | nr | 0.14 | 2.10 | 2.70 | 0.72 | 5.52 |
| 2.0m long | nr | 0.16 | 2.40 | 3.24 | 0.85 | 6.49 |
| 2.4m long | nr | 0.18 | 2.70 | 3.74 | 0.97 | 7.41 |
| Cleft chestnut tree stakes driven into the ground | | | | | | |
| 50-70mm wide | | | | | | |
| 1.7m long | nr | 0.10 | 1.50 | 1.77 | 0.49 | 3.76 |
| 2.0m long | nr | 0.12 | 1.80 | 2.15 | 0.59 | 4.54 |
| 2.4m long | nr | 0.14 | 2.10 | 2.57 | 0.70 | 5.37 |
| Tree ties, buckle type | | | | | | |
| 25 x 450mm | nr | 0.05 | 0.75 | 0.49 | 0.19 | 1.43 |
| 25 x 600mm | nr | 0.05 | 0.75 | 0.60 | 0.20 | 1.55 |
| Galvanised wire net tree guards | | | | | | |
| 150mm diameter x 450mm high | nr | 0.30 | 4.50 | 8.21 | 1.91 | 14.62 |
| 150mm diameter x 600mm high | nr | 0.30 | 4.50 | 9.39 | 2.08 | 15.97 |

| | | Unit | Plant £ | Mat'ls £ | O & P £ | Total £ |
|---|---|---|---|---|---|---|
| 200mm diameter x 1000mm high | nr | 0.35 | 4.90 | 9.66 | 2.18 | 16.74 |
| 200mm diameter x 1200mm high | nr | 0.35 | 4.90 | 10.16 | 2.26 | 17.32 |
| 250mm diameter x 1400mm high | nr | 0.35 | 4.90 | 11.48 | 2.46 | 18.84 |
| 250mm diameter x 1800mm high | nr | 0.35 | 4.90 | 12.93 | 2.67 | 20.50 |
| Bio-degradable tree guards | | | | | | |
| 600mm high | nr | 0.10 | 1.40 | 1.07 | 0.37 | 2.84 |
| 750mm high | nr | 0.10 | 1.40 | 1.25 | 0.40 | 3.05 |
| 900mm high | nr | 0.10 | 1.40 | 1.44 | 0.43 | 3.27 |
| 1200mm high | nr | 0.10 | 1.40 | 1.61 | 0.45 | 3.46 |

# Part Two

## UNIT RATES

Hard landscaping

Brick walling

Masonry

Precast concrete

Sub-bases

Beds and pavings

Fencing

Drainage

| | Unit | Labour | Hours £ | Mat'ls £ | O & P £ | Total £ |
|---|---|---|---|---|---|---|
| **BRICK WALLING** | | | | | | |
| Common bricks (basic price £140 per thousand) in cement mortar (1:3), stretcher bond | | | | | | |
| walls | | | | | | |
| half brick thick | m2 | 1.70 | 25.50 | 11.03 | 5.48 | 42.01 |
| half brick thick curved | m2 | 2.30 | 34.50 | 11.03 | 6.83 | 52.36 |
| one brick thick | m2 | 2.80 | 42.00 | 22.91 | 9.74 | 74.65 |
| one brick thick curved | m2 | 3.40 | 51.00 | 22.91 | 11.09 | 85.00 |
| one and a half brick thick | m2 | 3.50 | 52.50 | 35.53 | 13.20 | 101.23 |
| two brick thick | m2 | 4.20 | 63.00 | 48.12 | 16.67 | 127.79 |
| two brick thick battered | m2 | 4.80 | 72.00 | 48.12 | 18.02 | 138.14 |
| walls, facework one side | | | | | | |
| half brick thick | m2 | 1.80 | 27.00 | 11.03 | 5.70 | 43.73 |
| half brick thick curved | m2 | 2.40 | 36.00 | 11.03 | 7.05 | 54.08 |
| one brick thick | m2 | 2.90 | 43.50 | 22.91 | 9.96 | 76.37 |
| one brick thick curved | m2 | 3.50 | 52.50 | 22.91 | 11.31 | 86.72 |
| one and a half brick thick | m2 | 3.60 | 54.00 | 35.53 | 13.43 | 102.96 |
| two brick thick | m2 | 4.30 | 64.50 | 48.12 | 16.89 | 129.51 |
| two brick thick battered | m2 | 4.90 | 73.50 | 48.12 | 18.24 | 139.86 |
| walls, facework both sides | | | | | | |
| half brick thick | m2 | 1.90 | 28.50 | 11.03 | 5.93 | 45.46 |
| half brick thick curved | m2 | 2.50 | 37.50 | 11.03 | 7.28 | 55.81 |
| one brick thick | | 3.00 | 45.00 | 22.91 | 10.19 | 78.10 |
| one brick thick curved | m2 | 3.60 | 54.00 | 22.91 | 11.54 | 88.45 |
| one and a half brick thick | m2 | 3.70 | 55.50 | 22.91 | 11.76 | 90.17 |

| | Unit | Labour | Hours £ | Mat'ls £ | O & P £ | Total £ |
|---|---|---|---|---|---|---|
| **Common bricks (cont'd)** | | | | | | |
| two brick thick | m2 | 4.40 | 66.00 | 48.12 | 17.12 | 131.24 |
| two brick thick battered | m2 | 5.00 | 75.00 | 48.12 | 18.47 | 141.59 |
| Common bricks (basic price £200 per thousand) in cement mortar (1:3), stretcher bond | | | | | | |
| walls | | | | | | |
| half brick thick | m2 | 1.70 | 25.50 | 14.81 | 6.05 | 46.36 |
| half brick thick curved | m2 | 2.30 | 34.50 | 14.81 | 7.40 | 56.71 |
| one brick thick | m2 | 2.80 | 42.00 | 30.63 | 10.89 | 83.52 |
| one brick thick curved | m2 | 3.40 | 51.00 | 30.63 | 7.65 | 58.65 |
| one and a half brick thick | m2 | 3.50 | 52.50 | 46.74 | 14.89 | 114.13 |
| two brick thick | m2 | 4.20 | 63.00 | 46.74 | 16.46 | 126.20 |
| two brick thick battered | m2 | 4.80 | 72.00 | 63.11 | 20.27 | 155.38 |
| walls, facework one side | | | | | | |
| half brick thick | m2 | 1.80 | 27.00 | 14.81 | 6.27 | 48.08 |
| half brick thick curved | m2 | 2.40 | 36.00 | 14.81 | 7.62 | 58.43 |
| one brick thick | m2 | 2.90 | 43.50 | 30.63 | 11.12 | 85.25 |
| one brick thick curved | m2 | 3.50 | 52.50 | 30.63 | 12.47 | 95.60 |
| one and a half brick thick | m2 | 3.60 | 54.00 | 46.74 | 15.11 | 115.85 |
| two brick thick | m2 | 4.30 | 64.50 | 46.74 | 16.69 | 127.93 |
| two brick thick battered | m2 | 4.90 | 73.50 | 63.11 | 20.49 | 157.10 |
| walls, facework both sides | | | | | | |
| half brick thick | m2 | 1.90 | 28.50 | 14.81 | 6.50 | 49.81 |

| | Unit | Labour | Hours £ | Mat'ls £ | O & P £ | Total £ |
|---|---|---|---|---|---|---|
| half brick thick curved | m2 | 2.50 | 37.50 | 14.81 | 7.85 | 60.16 |
| one brick thick | m2 | 3.00 | 45.00 | 14.81 | 8.97 | 68.78 |
| one brick thick curved | m2 | 3.60 | 54.00 | 30.63 | 12.69 | 97.32 |
| one and a half brick thick | m2 | 3.70 | 55.50 | 30.63 | 12.92 | 99.05 |
| two brick thick | m2 | 4.40 | 66.00 | 46.74 | 16.91 | 129.65 |
| two brick thick battered | m2 | 5.00 | 75.00 | 63.11 | 20.72 | 158.83 |
| Facing bricks (basic price £250 per thousand) in cement mortar (1:3), English garden wall bond | | | | | | |
| walls, facework one side | | | | | | |
| one brick thick | m2 | 2.90 | 43.50 | 30.63 | 11.12 | 85.25 |
| one brick thick curved | m2 | 3.50 | 52.50 | 30.63 | 12.47 | 95.60 |
| one and a half brick thick | m2 | 3.60 | 54.00 | 46.74 | 15.11 | 115.85 |
| two brick thick | m2 | 4.30 | 64.50 | 63.11 | 19.14 | 146.75 |
| two brick thick battered | m2 | 4.90 | 73.50 | 63.11 | 20.49 | 157.10 |
| walls, facework both sides | | | | | | |
| one brick thick | m2 | 3.00 | 45.00 | 30.63 | 11.34 | 86.97 |
| one brick thick curved | m2 | 3.60 | 54.00 | 30.63 | 12.69 | 97.32 |
| one and a half brick thick | m2 | 3.70 | 55.50 | 46.74 | 15.34 | 117.58 |
| two brick thick | m2 | 4.40 | 66.00 | 63.11 | 19.37 | 148.48 |
| two brick thick battered | m2 | 5.00 | 75.00 | 63.11 | 20.72 | 158.83 |
| Isolated casings, half brick thick, faced all round | m2 | 2.30 | 34.50 | 18.51 | 7.95 | 60.96 |

| | Unit | Labour | Hours £ | Mat'ls £ | O & P £ | Total £ |
|---|---|---|---|---|---|---|
| Isolated casings, half brick thick, faced all round | m2 | 2.30 | 34.50 | 18.51 | 7.95 | 60.96 |
| Isolated piers, one brick thick, faced all round | m2 | 2.30 | 34.50 | 38.29 | 10.92 | 83.71 |
| Arches | | | | | | |
| 215mm high, 102mm wide | m2 | 1.30 | 19.50 | 4.80 | 3.65 | 27.95 |
| 215mm high, 215mm wide | m2 | 2.00 | 30.00 | 6.16 | 5.42 | 41.58 |
| 215mm high, 102mm wide, segmental | m2 | 2.70 | 40.50 | 9.13 | 7.44 | 57.07 |
| 215mm high, 315mm wide, segmental | m2 | 3.80 | 57.00 | 13.20 | 10.53 | 80.73 |
| Facing bricks (basic price £350 per thousand) in cement mortar (1:3), English garden wall bond | | | | | | |
| walls, facework one side | | | | | | |
| one brick thick | m2 | 2.90 | 43.50 | 49.22 | 13.91 | 106.63 |
| one brick thick curved | m2 | 3.50 | 52.50 | 49.22 | 15.26 | 116.98 |
| one and a half brick thick | m2 | 3.60 | 54.00 | 74.78 | 19.32 | 148.10 |
| two brick thick | m2 | 4.30 | 64.50 | 100.60 | 24.77 | 189.87 |
| two brick thick battered | m2 | 4.90 | 73.50 | 100.60 | 26.12 | 200.22 |
| walls, facework both sides | | | | | | |
| one brick thick | m2 | 3.00 | 45.00 | 49.22 | 14.13 | 108.35 |

| | Unit | Hours | Hours £ | Mat'ls £ | O & P £ | Total £ |
|---|---|---|---|---|---|---|
| one brick thick curved | m2 | 3.60 | 54.00 | 49.22 | 15.48 | 118.70 |
| one and a half brick thick | m2 | 3.70 | 55.50 | 49.22 | 15.71 | 120.43 |
| two brick thick | m2 | 4.40 | 66.00 | 74.78 | 21.12 | 161.90 |
| two brick thick battered | m2 | 5.00 | 75.00 | 100.60 | 26.34 | 201.94 |
| Isolated casings, half brick thick, faced all round | m2 | 2.30 | 34.50 | 30.33 | 9.72 | 74.55 |
| Isolated piers, one brick thick, faced all round | m2 | 2.30 | 34.50 | 61.52 | 14.40 | 110.42 |
| Arches | | | | | | |
| 215mm high, 102mm wide | m2 | 1.30 | 19.50 | 6.93 | 3.96 | 30.39 |
| 215mm high, 215mm wide | m2 | 2.00 | 30.00 | 7.30 | 5.60 | 42.90 |
| 215mm high, 102mm wide, segmental | m2 | 2.70 | 40.50 | 8.31 | 7.32 | 56.13 |
| 215mm high, 315mm wide, segmental | m2 | 3.80 | 57.00 | 11.94 | 10.34 | 79.28 |
| Facing bricks (basic price £250 per thousand) in cement mortar (1:3), flemish bond | | | | | | |
| walls, facework one side | | | | | | |
| one brick thick | m2 | 2.90 | 43.50 | 36.71 | 12.03 | 92.24 |
| one brick thick curved | m2 | 3.50 | 52.50 | 36.71 | 13.38 | 102.59 |
| one and a half brick thick | m2 | 3.60 | 54.00 | 56.19 | 16.53 | 126.72 |
| two brick thick | m2 | 4.30 | 64.50 | 65.61 | 19.52 | 149.63 |
| two brick thick battered | m2 | 4.90 | 73.50 | 65.61 | 20.87 | 159.98 |

| | Unit | Labour | Hours £ | Mat'ls £ | O & P £ | Total £ |
|---|---|---|---|---|---|---|
| walls, facework both sides | | | | | | |
| one brick thick | m2 | 3.00 | 45.00 | 36.71 | 12.26 | 93.97 |
| one brick thick curved | m2 | 3.60 | 54.00 | 36.71 | 13.61 | 104.32 |
| one and a half brick thick | m2 | 3.70 | 55.50 | 56.19 | 16.75 | 128.44 |
| two brick thick | m2 | 4.40 | 66.00 | 65.61 | 19.74 | 151.35 |
| two brick thick battered | m2 | 5.00 | 75.00 | 65.61 | 21.09 | 161.70 |
| Isolated casings, half brick thick, faced all round | m2 | 2.30 | 34.50 | 23.88 | 8.76 | 67.14 |
| Isolated piers, one brick thick, faced all round | m2 | 2.30 | 34.50 | 31.77 | 9.94 | 76.21 |
| Arches | | | | | | |
| 215mm high, 102mm wide | m2 | 1.30 | 19.50 | 4.80 | 3.65 | 27.95 |
| 215mm high, 215mm wide | m2 | 2.00 | 30.00 | 8.07 | 5.71 | 43.78 |
| 215mm high, 102mm wide, segmental | m2 | 2.70 | 40.50 | 9.13 | 7.44 | 57.07 |
| 215mm high, 315mm wide, segmental | m2 | 3.80 | 57.00 | 13.20 | 10.53 | 80.73 |
| Facing bricks (basic price £350 per thousand) in cement mortar (1:3), flemish bond | | | | | | |
| walls, facework one side | | | | | | |
| one brick thick | m2 | 2.90 | 43.50 | 24.26 | 10.16 | 77.92 |
| one brick thick curved | m2 | 3.50 | 52.50 | 24.26 | 11.51 | 88.27 |
| one and a half brick thick | m2 | 3.60 | 54.00 | 49.22 | 15.48 | 118.70 |

| | Unit | Labour | Hours £ | Mat'ls £ | O & P £ | Total £ |
|---|---|---|---|---|---|---|
| two brick thick | m2 | 4.30 | 64.50 | 74.78 | 20.89 | 160.17 |
| two brick thick battered | m2 | 4.90 | 73.50 | 100.60 | 26.12 | 200.22 |
| walls, facework both sides | | | | | | |
| one brick thick | m2 | 3.00 | 45.00 | 24.26 | 10.39 | 79.65 |
| one brick thick curved | m2 | 3.60 | 54.00 | 24.26 | 11.74 | 90.00 |
| one and a half brick thick | m2 | 3.70 | 55.50 | 49.22 | 15.71 | 120.43 |
| two brick thick | m2 | 4.40 | 66.00 | 74.78 | 21.12 | 161.90 |
| two brick thick battered | m2 | 5.00 | 75.00 | 100.60 | 26.34 | 201.94 |
| Isolated casings, half brick thick, faced all round | m2 | 2.30 | 34.50 | 26.67 | 9.18 | 70.35 |
| Isolated piers, one brick thick, faced all round | m2 | 2.30 | 34.50 | 54.14 | 13.30 | 101.94 |
| Arches | | | | | | |
| 215mm high, 102mm wide | m2 | 1.30 | 19.50 | 6.93 | 3.96 | 30.39 |
| 215mm high, 215mm wide | m2 | 2.00 | 30.00 | 7.30 | 5.60 | 42.90 |
| 215mm high, 102mm wide, segmental | m2 | 2.70 | 40.50 | 8.27 | 7.32 | 56.09 |
| 215mm high, 315mm wide, segmental | m2 | 3.80 | 57.00 | 11.88 | 10.33 | 79.21 |

| | Unit | Labour | Hours £ | Mat'ls £ | O & P £ | Total £ |
|---|---|---|---|---|---|---|
| **MASONRY** | | | | | | |
| Random rubble walling, laid dry, thickness | | | | | | |
| 300mm | m2 | 3.00 | 45.00 | 50.38 | 14.31 | 109.69 |
| 450mm | m2 | 3.50 | 52.50 | 75.52 | 19.20 | 147.22 |
| 500mm | m2 | 3.75 | 56.25 | 85.20 | 21.22 | 162.67 |
| Random rubble walling, laid dry, battered one side, thickness | | | | | | |
| 300mm | m2 | 3.25 | 48.75 | 54.63 | 15.51 | 118.89 |
| 450mm | m2 | 3.75 | 56.25 | 81.87 | 20.72 | 158.84 |
| 500mm | m2 | 4.00 | 60.00 | 92.36 | 22.85 | 175.21 |
| Random rubble walling, laid dry, battered both sides, thickness | | | | | | |
| 300mm | m2 | 3.50 | 52.50 | 54.63 | 16.07 | 123.20 |
| 450mm | m2 | 4.00 | 60.00 | 81.87 | 21.28 | 163.15 |
| 500mm | m2 | 4.25 | 63.75 | 92.36 | 23.42 | 179.53 |
| Random rubble walling, laid in gauged mortar (1:1:6), thickness | | | | | | |
| 300mm | m2 | 3.20 | 48.00 | 70.23 | 17.73 | 135.96 |
| 450mm | m2 | 3.70 | 55.50 | 102.67 | 23.73 | 181.90 |
| 500mm | m2 | 3.90 | 58.50 | 115.76 | 26.14 | 200.40 |

| | Unit | Labour | Hours £ | Mat'ls £ | O & P £ | Total £ |
|---|---|---|---|---|---|---|
| Random rubble walling, laid in gauged mortar (1:1:6), thickness, battered one side | | | | | | |
| 300mm | m2 | 3.45 | 51.75 | 70.23 | 18.30 | 140.28 |
| 450mm | m2 | 4.00 | 60.00 | 102.67 | 24.40 | 187.07 |
| 500mm | m2 | 4.15 | 62.25 | 115.76 | 26.70 | 204.71 |
| Random rubble walling, laid in gauged mortar (1:1:6), thickness, battered both sides | | | | | | |
| 300mm | m2 | 3.95 | 59.25 | 70.23 | 19.42 | 148.90 |
| 450mm | m2 | 4.50 | 67.50 | 102.67 | 25.53 | 195.70 |
| 500mm | m2 | 4.65 | 69.75 | 115.76 | 27.83 | 213.34 |
| Irregular coursed rubble walling, laid in gauged mortar (1:1:6), thickness | | | | | | |
| 300mm | m2 | 2.10 | 31.50 | 70.23 | 15.26 | 116.99 |
| 450mm | m2 | 2.90 | 43.50 | 102.67 | 21.93 | 168.10 |
| 500mm | m2 | 3.40 | 51.00 | 115.76 | 25.01 | 191.77 |
| Coursed rubble walling, laid in gauged mortar (1:1:6), thickness | | | | | | |
| 300mm | m2 | 2.40 | 36.00 | 70.23 | 15.93 | 122.16 |
| 450mm | m2 | 3.20 | 48.00 | 102.67 | 22.60 | 173.27 |
| 500mm | m2 | 3.70 | 55.50 | 115.76 | 25.69 | 196.95 |

| | Unit | Labour | Hours £ | Mat'ls £ | O & P £ | Total £ |
|---|---|---|---|---|---|---|
| Fair raking cutting on stone walling, thickness | | | | | | |
| 300mm | m | 0.80 | 12.00 | 3.17 | 2.28 | 17.45 |
| 450mm | m | 1.20 | 18.00 | 5.13 | 3.47 | 26.60 |
| 500mm | m | 1.80 | 27.00 | 5.89 | 4.93 | 37.82 |
| Form level bed on stone walling, thickness | | | | | | |
| 300mm | m | 0.20 | 3.00 | 0.90 | 0.59 | 4.49 |
| 450mm | m | 0.35 | 5.25 | 1.45 | 1.01 | 7.71 |
| 500mm | m | 0.50 | 7.50 | 1.58 | 1.36 | 10.44 |
| Sundries | | | | | | |
| Form holes for pipes up to 50mm diameter through stone walling, thickness | | | | | | |
| 300mm | m | 1.20 | 18.00 | 0.00 | 2.70 | 20.70 |
| 450mm | m | 1.80 | 27.00 | 0.00 | 4.05 | 31.05 |
| 500mm | m | 2.65 | 39.75 | 0.00 | 5.96 | 45.71 |
| Form holes for pipes 50–100mm diameter through stone walling, thickness | | | | | | |
| 300mm | m | 1.40 | 21.00 | 0.00 | 3.15 | 24.15 |
| 450mm | m | 2.00 | 30.00 | 0.00 | 4.50 | 34.50 |
| 500mm | m | 2.85 | 42.75 | 0.00 | 6.41 | 49.16 |

| | Unit | Labour | Hours £ | Mat'ls £ | O & P £ | Total £ |
|---|---|---|---|---|---|---|
| Form holes for pipes over 100mm diameter through stone walling, thickness | | | | | | |
| 300mm | m | 2.40 | 36.00 | 0.00 | 5.40 | 41.40 |
| 450mm | m | 3.00 | 45.00 | 0.00 | 6.75 | 51.75 |
| 500mm | m | 3.85 | 57.75 | 0.00 | 8.66 | 66.41 |
| Build in ends of steel sections in stone walling, size | | | | | | |
| not exceeding 250mm deep | m | 0.70 | 10.50 | 0.00 | 1.58 | 12.08 |
| 250–500mm deep | m | 0.90 | 13.50 | 0.00 | 2.03 | 15.53 |
| over 500mm deep | m | 1.10 | 16.50 | 0.00 | 2.48 | 18.98 |
| Mortices in stone walling, size | | | | | | |
| 50 × 50 × 100mm | nr | 0.60 | 9.00 | 0.00 | 1.35 | 10.35 |
| 50 × 50 × 150mm | nr | 0.65 | 9.75 | 0.00 | 1.46 | 11.21 |
| 75 × 75 × 100mm | nr | 0.70 | 10.50 | 0.00 | 1.58 | 12.08 |
| 75 × 75 × 150mm | nr | 0.75 | 11.25 | 0.00 | 1.69 | 12.94 |
| Grout up mortices in cement mortar (1:3) | | | | | | |
| 50 × 50 × 100mm | nr | 0.10 | 1.50 | 0.05 | 0.23 | 1.78 |
| 50 × 50 × 150mm | nr | 0.15 | 2.25 | 0.22 | 0.37 | 2.84 |
| 75 × 75 × 100mm | nr | 0.15 | 2.25 | 0.33 | 0.39 | 2.97 |
| 75 × 75 × 150mm | nr | 0.20 | 3.00 | 0.38 | 0.51 | 3.89 |

| | Unit | Labour | Hours £ | Mat'ls £ | O & P £ | Total £ |
|---|---|---|---|---|---|---|
| **Damp proof courses** | | | | | | |
| Hessian based bitumen damp-proof course bedded in gauged mortar(1:1:6) | | | | | | |
| horizontal, width | | | | | | |
| over 215mm | m2 | 0.35 | 5.25 | 2.32 | 1.14 | 8.71 |
| 112mm | m | 0.05 | 0.75 | 1.14 | 0.28 | 2.17 |
| vertical, width | | | | | | |
| over 215mm | m2 | 0.40 | 6.00 | 2.32 | 1.25 | 9.57 |
| 112mm | m | 0.07 | 1.05 | 1.14 | 0.33 | 2.52 |
| Fibre-based bitumen damp-proof course bedded in gauged mortar(1:1:6) | | | | | | |
| horizontal, width | | | | | | |
| over 215mm | m2 | 0.40 | 6.00 | 2.16 | 1.22 | 9.38 |
| 112mm | m | 0.05 | 0.75 | 0.98 | 0.26 | 1.99 |
| vertical, width | | | | | | |
| over 215mm | m2 | 0.45 | 6.75 | 2.16 | 1.34 | 10.25 |
| 112mm | m | 0.07 | 1.05 | 0.98 | 0.30 | 2.33 |
| Pitch polymer damp proof course in gauged mortar (1:1:6) | | | | | | |
| horizontal, width | | | | | | |
| over 215mm | m2 | 0.40 | 6.00 | 2.22 | 1.23 | 9.45 |
| 112mm | m | 0.05 | 0.75 | 1.12 | 0.28 | 2.15 |
| vertical, width | | | | | | |
| over 215mm | m2 | 0.45 | 6.75 | 2.22 | 1.35 | 10.32 |
| 112mm | m | 0.07 | 1.05 | 1.12 | 0.33 | 2.50 |

| | Unit | Labour | Hours £ | Mat'ls £ | O & P £ | Total £ |
|---|---|---|---|---|---|---|
| Two courses of slates bedded in cement mortar (1:3) | | | | | | |
| horizontal, width | | | | | | |
| over 225mm | m2 | 0.87 | 13.05 | 17.45 | 4.58 | 35.08 |
| 112mm | m | 0.33 | 4.95 | 2.32 | 1.09 | 8.36 |
| vertical, width | | | | | | |
| over 225mm | m2 | 0.95 | 14.25 | 17.45 | 4.76 | 36.46 |
| 112mm | m | 0.35 | 5.25 | 2.32 | 1.14 | 8.71 |
| **PRECAST CONCRETE** | | | | | | |
| Precast concrete copings twice weatherted and twice throated, size | | | | | | |
| 75 × 150 × 900mm | nr | 0.30 | 4.50 | 7.12 | 1.74 | 13.36 |
| 75 × 300 × 900mm | nr | 0.40 | 6.00 | 8.71 | 2.21 | 16.92 |
| Precast concrete pier caps, four times weathered, size | | | | | | |
| 300 × 300 × 75mm | nr | 1.00 | 15.00 | 7.04 | 3.31 | 25.35 |
| 450 × 450 × 75mm | nr | 1.20 | 18.00 | 9.90 | 4.19 | 32.09 |
| **Kerbs and edgings** | | | | | | |
| Excavate trench by hand for kerb foundation | | | | | | |
| 200 × 75mm | m | 0.10 | 1.50 | 0.00 | 0.23 | 1.73 |
| 250 × 100mm | m | 0.12 | 1.80 | 0.00 | 0.27 | 2.07 |
| 300 × 100mm | m | 0.14 | 2.10 | 0.00 | 0.32 | 2.42 |
| 450 × 150mm | m | 0.20 | 3.00 | 0.00 | 0.45 | 3.45 |
| 600 × 200mm | m | 0.40 | 6.00 | 0.00 | 0.90 | 6.90 |

| | Unit | Labour | Hours £ | Mat'ls £ | O & P £ | Total £ |
|---|---|---|---|---|---|---|
| Excavate curved trench by hand for kerb foundation | | | | | | |
| 250 × 100mm | m | 0.12 | 1.80 | 0.00 | 0.27 | 2.07 |
| 250 × 100mm | m | 0.14 | 2.10 | 0.00 | 0.32 | 2.42 |
| 300 × 100mm | m | 0.16 | 2.40 | 0.00 | 0.36 | 2.76 |
| 600 × 200mm | m | 0.24 | 3.60 | 0.00 | 0.54 | 4.14 |
| 600 × 200mm | m | 0.44 | 6.60 | 0.00 | 0.99 | 7.59 |
| Site-mixed concrete in foundation for kerb size | | | | | | |
| 200 × 75mm | m | 0.03 | 0.45 | 1.19 | 0.25 | 1.89 |
| 250 × 100mm | m | 0.04 | 0.60 | 1.98 | 0.39 | 2.97 |
| 300 × 100mm | m | 0.05 | 0.75 | 2.38 | 0.47 | 3.60 |
| 450 × 150mm | m | 0.18 | 2.70 | 5.15 | 1.18 | 9.03 |
| 600 × 200mm | m | 0.30 | 4.50 | 9.36 | 2.08 | 15.94 |
| Precast concrete kerbs, channels and edgings, jointed and pointed in cement mortar | | | | | | |
| kerbs, straight | | | | | | |
| 127 × 254mm | m | 0.40 | 6.00 | 6.68 | 1.90 | 14.58 |
| 152 × 305mm | m | 0.45 | 6.75 | 8.36 | 2.27 | 17.38 |
| kerbs, curved | | | | | | |
| 127 × 254mm | m | 0.50 | 7.50 | 8.00 | 2.33 | 17.83 |
| 152 × 305mm | m | 0.55 | 8.25 | 12.42 | 3.10 | 23.77 |
| channels, straight | | | | | | |
| 127 × 254mm | m | 0.40 | 6.00 | 7.91 | 2.09 | 16.00 |

| | Unit | Labour | Hours £ | Mat'ls £ | O & P £ | Total £ |
|---|---|---|---|---|---|---|
| channels, curved | | | | | | |
| 127 × 254mm | m | 0.50 | 7.50 | 9.10 | 2.49 | 19.09 |
| edgings, straight | | | | | | |
| 51 × 152mm | m | 0.30 | 4.50 | 3.47 | 1.20 | 9.17 |
| 51 × 203mm | m | 0.30 | 4.50 | 4.15 | 1.30 | 9.95 |

**SUB-BASES**

Beds and bases compacting in layers and grading

| | Unit | Labour | Hours £ | Mat'ls £ | O & P £ | Total £ |
|---|---|---|---|---|---|---|
| average thickness, 100mm | | | | | | |
| granular fill | m2 | 0.10 | 1.50 | 3.31 | 0.72 | 5.53 |
| sand | m2 | 0.12 | 1.80 | 3.14 | 0.74 | 5.68 |
| hardcore | m2 | 0.13 | 1.95 | 3.02 | 0.75 | 5.72 |
| average thickness, 150mm | | | | | | |
| granular fill | m2 | 0.12 | 1.80 | 4.15 | 0.89 | 6.84 |
| sand | m2 | 0.14 | 2.10 | 3.98 | 0.91 | 6.99 |
| hardcore | m2 | 0.16 | 2.40 | 3.91 | 0.95 | 7.26 |
| average thickness, 200mm | | | | | | |
| granular fill | m2 | 0.14 | 2.10 | 5.38 | 1.12 | 8.60 |
| sand | m2 | 0.16 | 2.40 | 4.79 | 1.08 | 8.27 |
| hardcore | m2 | 0.18 | 2.70 | 4.71 | 1.11 | 8.52 |

| | Unit | Labour | Hours £ | Mat'ls £ | O & P £ | Total £ |
|---|---|---|---|---|---|---|
| **BEDS AND PAVINGS** | | | | | | |
| **Preparatory work** | | | | | | |
| Excavate by hand to form path, depth | | | | | | |
| 100mm | m2 | 0.25 | 3.75 | 0.00 | 0.56 | 4.31 |
| 150mm | m2 | 0.35 | 5.25 | 0.00 | 0.79 | 6.04 |
| 200mm | m2 | 0.45 | 6.75 | 0.00 | 1.01 | 7.76 |
| 250mm | m2 | 0.60 | 9.00 | 0.00 | 1.35 | 10.35 |
| 300mm | m2 | 0.75 | 11.25 | 0.00 | 1.69 | 12.94 |
| Hardcore bed compacted in layers and blinded with sand | | | | | | |
| 100mm | m2 | 0.08 | 1.20 | 3.02 | 0.63 | 4.85 |
| 150mm | m2 | 0.10 | 1.50 | 3.91 | 0.81 | 6.22 |
| 200mm | m2 | 0.12 | 1.80 | 4.71 | 0.98 | 7.49 |
| **Concrete beds** | | | | | | |
| Site-mixed concrete in beds | | | | | | |
| not exceeding 150mm thick | m3 | 2.80 | 42.00 | 87.33 | 19.40 | 148.73 |
| 150 to 450mm thick | m3 | 2.20 | 33.00 | 87.33 | 18.05 | 138.38 |
| Formwork to sides of concrete bases | | | | | | |
| not exceeding 250mm wide | m | 0.65 | 9.75 | 2.05 | 1.77 | 13.57 |
| 250 to 500mm wide | m | 0.90 | 13.50 | 3.39 | 2.53 | 19.42 |

| | Unit | Labour | Hours £ | Mat'ls £ | O & P £ | Total £ |
|---|---|---|---|---|---|---|
| **Concrete beds (cont'd)** | | | | | | |
| Steel fabric reinforcement laid in concrete beds | | | | | | |
| ref A142, 2.22kg/m2 | m2 | 0.15 | 2.25 | 1.56 | 0.57 | 4.38 |
| ref A193, 3.02kg/m2 | m2 | 0.18 | 2.70 | 1.89 | 0.69 | 5.28 |
| Expansion joint, impregnated fibre-base joint filler, formed joint | | | | | | |
| 12.5mm thick | | | | | | |
| not exceeding 150mm | m | 0.18 | 2.70 | 2.59 | 0.79 | 6.08 |
| 150 to 300mm wide | m | 0.24 | 3.60 | 4.10 | 1.16 | 8.86 |
| 300 to 450mm wide | m | 0.30 | 4.50 | 6.16 | 1.60 | 12.26 |
| 25mm thick | | | | | | |
| not exceeding 150mm | m | 0.20 | 3.00 | 3.68 | 1.00 | 7.68 |
| 150 to 300mm wide | m | 0.26 | 3.90 | 5.41 | 1.40 | 10.71 |
| 300 to 450mm wide | m | 0.32 | 4.80 | 8.37 | 1.98 | 15.15 |
| Treat surfaces of concrete before setting | | | | | | |
| tamping | m2 | 0.06 | 0.90 | 0.00 | 0.14 | 1.04 |
| floating | m2 | 0.10 | 1.50 | 0.00 | 0.23 | 1.73 |
| trowelling | m2 | 0.15 | 2.25 | 0.00 | 0.34 | 2.59 |
| **Granular pavings** | | | | | | |
| Washed pit gravel, 19mm, bed thickness | | | | | | |
| 50mm | m2 | 0.06 | 0.90 | 2.37 | 0.49 | 3.76 |
| 75mm | m2 | 0.08 | 1.20 | 3.49 | 0.70 | 5.39 |

| | Unit | Labour | Hours £ | Mat'ls £ | O & P £ | Total £ |
|---|---|---|---|---|---|---|
| Washed pit gravel, 38mm, bed thickness | | | | | | |
| 50mm | m2 | 0.06 | 0.90 | 2.19 | 0.46 | 3.55 |
| 75mm | m2 | 0.08 | 1.20 | 3.37 | 0.69 | 5.26 |
| Hoggin bed laid on hard-core, bed thickness | | | | | | |
| 100mm | m2 | 0.14 | 2.10 | 2.55 | 0.70 | 5.35 |
| 150mm | m2 | 0.18 | 2.70 | 3.86 | 0.98 | 7.54 |
| 200mm | m2 | 0.22 | 3.30 | 4.65 | 1.19 | 9.14 |
| Bark laid on prepared surfaces, bed thickness 100mm | m2 | 0.05 | 0.75 | 2.39 | 0.47 | 3.61 |
| Impregnated softwood edging boards with 38 × 38 × 225mm pegs at 1200mm centres | | | | | | |
| 38 × 100mm | m2 | 0.05 | 0.75 | 2.22 | 0.45 | 3.42 |
| 38 × 150mm | m2 | 0.07 | 1.05 | 2.58 | 0.54 | 4.17 |
| **Brick/block pavings** | | | | | | |
| Brick paving, £250 per 1000, 215 × 103 × 65mm, laid to falls and cross falls bedding in cement mortar 15mm thick | | | | | | |
| straight joints both ways | | | | | | |
| bricks laid flat | m2 | 1.00 | 15.00 | 25.77 | 6.12 | 46.89 |
| bricks laid on edge | m2 | 1.25 | 18.75 | 39.16 | 8.69 | 66.60 |

| | Unit | Labour | Hours £ | Mat'ls £ | O & P £ | Total £ |
|---|---|---|---|---|---|---|
| herringbone pattern | | | | | | |
| bricks laid flat | m2 | 1.25 | 18.75 | 25.77 | 6.68 | 51.20 |
| bricks laid on edge | m2 | 1.50 | 22.50 | 39.16 | 9.25 | 70.91 |
| Brick paving, £350 per 1000, 215 × 103 × 65mm, laid to falls and cross falls bedding in cement mortar 15mm thick | | | | | | |
| straight joints both ways | | | | | | |
| bricks laid flat | m2 | 1.00 | 15.00 | 30.87 | 6.88 | 52.75 |
| bricks laid on edge | m2 | 1.25 | 18.75 | 44.90 | 9.55 | 73.20 |
| herringbone pattern | | | | | | |
| bricks laid flat | m2 | 1.25 | 18.75 | 30.87 | 7.44 | 57.06 |
| bricks laid on edge | m2 | 1.50 | 22.50 | 44.90 | 10.11 | 77.51 |
| Brick paving, £500 per 1000, 215 × 103 × 65mm, laid to falls and cross falls bedding in cement mortar 15mm thick | | | | | | |
| straight joints both ways | | | | | | |
| bricks laid flat | m2 | 1.00 | 15.00 | 37.80 | 7.92 | 60.72 |
| bricks laid on edge | m2 | 1.25 | 18.75 | 51.81 | 10.58 | 81.14 |
| herringbone pattern | | | | | | |
| bricks laid flat | m2 | 1.25 | 18.75 | 37.80 | 8.48 | 65.03 |
| bricks laid on edge | m2 | 1.50 | 22.50 | 51.11 | 11.04 | 84.65 |

| | Unit | Labour | Hours £ | Mat'ls £ | O & P £ | Total £ |
|---|---|---|---|---|---|---|
| **Precast concrete pavings** | | | | | | |
| Precast concrete paving flags, spot bedded in cement mortar, straight both ways, jointing in cement mortar brushed in | | | | | | |
| natural colour | | | | | | |
| 450 × 450 × 50mm | m2 | 0.55 | 8.25 | 16.81 | 3.76 | 28.82 |
| 450 × 600 × 50mm | m2 | 0.55 | 8.25 | 13.54 | 3.27 | 25.06 |
| 600 × 600 × 50mm | m2 | 0.50 | 7.50 | 11.34 | 2.83 | 21.67 |
| 600 × 900 × 50mm | m2 | 0.45 | 6.75 | 10.60 | 2.60 | 19.95 |
| coloured | | | | | | |
| 450 × 450 × 50mm | m2 | 0.55 | 8.25 | 20.38 | 4.29 | 32.92 |
| 450 × 600 × 50mm | m2 | 0.55 | 8.25 | 17.11 | 3.80 | 29.16 |
| 600 × 600 × 50mm | m2 | 0.50 | 7.50 | 15.82 | 3.50 | 26.82 |
| 600 × 900 × 50mm | m2 | 0.45 | 6.75 | 13.98 | 3.11 | 23.84 |
| Reconstituted York stone paving flags, spot bedded in cement mortar, straight both ways, jointing in cement mortar brushed in, 45mm thick, size | | | | | | |
| 300 x 300mm | m2 | 1.00 | 15.00 | 28.89 | 6.58 | 50.47 |
| 300 x 450mm | m2 | 0.80 | 12.00 | 27.45 | 5.92 | 45.37 |
| 300 x 600mm | m2 | 0.60 | 9.00 | 26.50 | 5.33 | 40.83 |
| 450 x 450mm | m2 | 0.55 | 8.25 | 24.99 | 4.99 | 38.23 |
| 600 x 600mm | m2 | 0.50 | 7.50 | 25.81 | 5.00 | 38.31 |

| | Unit | Labour | Hours £ | Mat'ls £ | O & P £ | Total £ |
|---|---|---|---|---|---|---|
| York stone paving flags spot bedded in cement mortar, straight both ways, jointing in cement mortar brushed in | | | | | | |
| 50mm thick | | | | | | |
| cut to size | m2 | 0.80 | 12.00 | 95.52 | 16.13 | 123.65 |
| random sizes | m2 | 0.90 | 13.50 | 79.90 | 14.01 | 107.41 |
| 63mm thick | | | | | | |
| cut to size | m2 | 0.85 | 12.75 | 105.11 | 17.68 | 135.54 |
| random sizes | m2 | 0.95 | 14.25 | 89.31 | 15.53 | 119.09 |
| 75mm thick | | | | | | |
| cut to size | m2 | 0.90 | 13.50 | 118.77 | 19.84 | 152.11 |
| random sizes | m2 | 1.00 | 15.00 | 99.90 | 17.24 | 132.14 |
| Natural granite setts bedded in cement mortar, size 100 × 100 × 150mm | m2 | 2.75 | 41.25 | 35.82 | 11.56 | 88.63 |
| Concrete setts bedded in cement mortar, size | | | | | | |
| 150 × 150 × 80mm | m2 | 2.75 | 41.25 | 33.87 | 11.27 | 86.39 |
| 200 × 150 × 80mm | m2 | 2.50 | 37.50 | 33.87 | 10.71 | 82.08 |
| Concrete paviors, bedded in sand, with sand brushed in, size | | | | | | |
| 200 × 100 × 60mm | m2 | 1.00 | 15.00 | 11.55 | 3.98 | 30.53 |
| 200 × 100 × 80mm | m2 | 1.00 | 15.00 | 12.80 | 4.17 | 31.97 |

| | Unit | Labour | Hours £ | Mat'ls £ | O & P £ | Total £ |
|---|---|---|---|---|---|---|
| **FENCING** | | | | | | |
| **Fencing posts** | | | | | | |
| Galvanised steel posts, size 40 × 40 × 5mm, driven into ground | | | | | | |
| 900mm high | nr | 0.05 | 0.75 | 1.83 | 0.39 | 2.97 |
| 1200mm high | nr | 0.05 | 0.75 | 2.18 | 0.44 | 3.37 |
| 1400mm high | nr | 0.10 | 1.50 | 2.48 | 0.60 | 4.58 |
| 1800mm high | nr | 0.15 | 2.25 | 3.08 | 0.80 | 6.13 |
| 2400mm high | nr | 0.15 | 2.25 | 3.49 | 0.86 | 6.60 |
| Galvanised steel posts, size 40 × 40 × 5mm, with one strut, set in concrete | | | | | | |
| 900mm high | nr | 0.20 | 3.00 | 20.98 | 3.60 | 27.58 |
| 1200mm high | nr | 0.20 | 3.00 | 24.77 | 4.17 | 31.94 |
| 1400mm high | nr | 0.25 | 3.75 | 27.90 | 4.75 | 36.40 |
| 1800mm high | nr | 0.30 | 4.50 | 31.87 | 5.46 | 41.83 |
| 2400mm high | nr | 0.30 | 4.50 | 36.94 | 6.22 | 47.66 |
| Galvanised steel posts, size 40 × 40 × 5mm, with two struts, driven into ground | | | | | | |
| 900mm high | nr | 0.30 | 4.50 | 39.51 | 6.60 | 50.61 |
| 1200mm high | nr | 0.30 | 4.50 | 43.31 | 7.17 | 54.98 |
| 1400mm high | nr | 0.35 | 5.25 | 45.90 | 7.67 | 58.82 |
| 1800mm high | nr | 0.40 | 6.00 | 49.08 | 8.26 | 63.34 |
| 2400mm high | nr | 0.40 | 6.00 | 55.95 | 9.29 | 71.24 |

| | Unit | Labour | Hours £ | Mat'ls £ | O & P £ | Total £ |
|---|---|---|---|---|---|---|
| Precast concrete posts size 100 × 100mm, set in concrete | | | | | | |
| 900mm high | nr | 0.25 | 3.75 | 20.98 | 3.71 | 28.44 |
| 1200mm high | nr | 0.25 | 3.75 | 24.77 | 4.28 | 32.80 |
| 1400mm high | nr | 0.30 | 4.50 | 27.90 | 4.86 | 37.26 |
| 1800mm high | nr | 0.35 | 5.25 | 31.87 | 5.57 | 42.69 |
| 2400mm high | nr | 0.35 | 5.25 | 36.94 | 6.33 | 48.52 |
| Precast concrete posts size 100 × 100mm, with one strut, set in concrete | | | | | | |
| 900mm high | nr | 0.40 | 6.00 | 46.17 | 7.83 | 60.00 |
| 1200mm high | nr | 0.40 | 6.00 | 49.97 | 8.40 | 64.37 |
| 1400mm high | nr | 0.45 | 6.75 | 59.14 | 9.88 | 75.77 |
| 1800mm high | nr | 0.55 | 8.25 | 63.17 | 10.71 | 82.13 |
| 2400mm high | nr | 0.55 | 8.25 | 83.04 | 13.69 | 104.98 |
| Precast concrete posts size 100 × 100mm, with two struts, set in concrete | | | | | | |
| 900mm high | nr | 0.50 | 7.50 | 63.61 | 10.67 | 81.78 |
| 1200mm high | nr | 0.50 | 7.50 | 68.42 | 11.39 | 87.31 |
| 1400mm high | nr | 0.55 | 8.25 | 78.34 | 12.99 | 99.58 |
| 1800mm high | nr | 0.60 | 9.00 | 90.47 | 14.92 | 114.39 |
| 2400mm high | nr | 0.60 | 9.00 | 96.37 | 15.81 | 121.18 |

| | Unit | Labour | Hours £ | Mat'ls £ | O & P £ | Total £ |
|---|---|---|---|---|---|---|
| **Chainlink fencing** | | | | | | |
| Chainlink fencing, galvanised steel mesh on three line, wires fixed to posts (not included) | | | | | | |
| 900mm high | m | 0.15 | 2.25 | 1.82 | 0.61 | 4.68 |
| 1200mm high | m | 0.15 | 2.25 | 1.90 | 0.62 | 4.77 |
| 1400mm high | m | 0.20 | 3.00 | 3.18 | 0.93 | 7.11 |
| 1800mm high | m | 0.25 | 3.75 | 3.83 | 1.14 | 8.72 |
| 2400mm high | m | 0.25 | 3.75 | 5.20 | 1.34 | 10.29 |
| Chainlink fencing, plastic coated steel mesh on three line, wires fixed to posts (not included) | | | | | | |
| 900mm high | m | 0.15 | 2.25 | 1.92 | 0.63 | 4.80 |
| 1200mm high | m | 0.15 | 2.25 | 2.00 | 0.64 | 4.89 |
| 1400mm high | m | 0.20 | 3.00 | 3.28 | 0.94 | 7.22 |
| 1800mm high | m | 0.25 | 3.75 | 3.93 | 1.15 | 8.83 |
| 2400mm high | m | 0.25 | 3.75 | 5.30 | 1.36 | 10.41 |
| **Chestnut fencing** | | | | | | |
| Chestnut fencing with pales at 75mm centres, fixed to 75 mm posts at 3 metre centres driven into ground | | | | | | |
| two line wire | | | | | | |
| 900mm high | m | 0.25 | 3.75 | 4.52 | 1.24 | 9.51 |
| 1100mm high | m | 0.30 | 4.50 | 4.78 | 1.39 | 10.67 |
| 1250mm high | m | 0.35 | 5.25 | 5.07 | 1.55 | 11.87 |

| | Unit | Labour | Hours £ | Mat'ls £ | O & P £ | Total £ |
|---|---|---|---|---|---|---|
| **Chestnut fencing (cont'd)** | | | | | | |
| three line wire | | | | | | |
| 900mm high | m | 0.25 | 3.75 | 4.83 | 1.29 | 9.87 |
| 1100mm high | m | 0.30 | 4.50 | 4.99 | 1.42 | 10.91 |
| 1250mm high | m | 0.35 | 5.25 | 5.23 | 1.57 | 12.05 |
| **Strained wire fencing** | | | | | | |
| Strained wire fencing fixed to concrete posts (not included) | | | | | | |
| 1000mm high, 3 wires | m | 0.10 | 1.50 | 0.48 | 0.30 | 2.28 |
| 1200mm high, 3 wires | m | 0.11 | 1.65 | 0.48 | 0.32 | 2.45 |
| 1500mm high, 4 wires | m | 0.12 | 1.80 | 0.72 | 0.38 | 2.90 |
| **Barbed wire fencing** | | | | | | |
| Barbed wire fencing fixed to concrete posts (not included) | | | | | | |
| 1000mm high, 3 wires | m | 0.14 | 2.10 | 0.61 | 0.41 | 3.12 |
| 1200mm high, 3 wires | m | 0.16 | 2.40 | 0.72 | 0.47 | 3.59 |
| 1500mm high, 4 wires | m | 0.18 | 2.70 | 0.84 | 0.53 | 4.07 |

| | Unit | Labour | Hours £ | Mat'ls £ | O & P £ | Total £ |
|---|---|---|---|---|---|---|
| **DRAINAGE** | | | | | | |
| Excavate drain trench by hand, support sides, level and ram bottom of trench, backfill and consolidate with excavated material and and remove surplus to tip for pipes 100mm and 150mm diameter, average depth of trench | | | | | | |
| 0.50m | m | 1.20 | 18.00 | 0.00 | 2.70 | 20.70 |
| 0.75m | m | 1.90 | 28.50 | 0.00 | 4.28 | 32.78 |
| 1.00m | m | 2.50 | 37.50 | 0.00 | 5.63 | 43.13 |
| 1.25m | m | 4.00 | 60.00 | 0.00 | 9.00 | 69.00 |
| 1.50m | m | 5.00 | 75.00 | 0.00 | 11.25 | 86.25 |
| 1.75m | m | 6.00 | 90.00 | 0.00 | 13.50 | 103.50 |
| 2.00m | m | 7.20 | 108.00 | 0.00 | 16.20 | 124.20 |
| 2.25m | m | 8.20 | 123.00 | 0.00 | 18.45 | 141.45 |
| 2.50m | m | 9.50 | 142.50 | 0.00 | 21.38 | 163.88 |
| 2.75m | m | 10.50 | 157.50 | 0.00 | 23.63 | 181.13 |
| 3.00m | m | 12.00 | 180.00 | 0.00 | 27.00 | 207.00 |
| Excavate drain trench by hand, support sides, level and ram bottom of trench, backfill and consolidate with excavated material and and remove surplus to tip for pipes 225mm diameter, average depth of trench | | | | | | |
| 0.50m | m | 1.30 | 19.50 | 0.00 | 2.93 | 22.43 |
| 0.75m | m | 2.10 | 31.50 | 0.00 | 4.73 | 36.23 |
| 1.00m | m | 2.70 | 40.50 | 0.00 | 6.08 | 46.58 |

| | Unit | Labour | Hours £ | Mat'ls £ | O & P £ | Total £ |
|---|---|---|---|---|---|---|
| **Excavate drain trench (cont'd)** | | | | | | |
| 1.25m | m | 4.30 | 64.50 | 0.00 | 9.68 | 74.18 |
| 1.50m | m | 5.40 | 81.00 | 0.00 | 12.15 | 93.15 |
| 1.75m | m | 6.60 | 99.00 | 0.00 | 14.85 | 113.85 |
| 2.00m | m | 7.10 | 106.50 | 0.00 | 15.98 | 122.48 |
| 2.25m | m | 9.00 | 135.00 | 0.00 | 20.25 | 155.25 |
| 2.50m | m | 10.20 | 153.00 | 0.00 | 22.95 | 175.95 |
| 2.75m | m | 11.30 | 169.50 | 0.00 | 25.43 | 194.93 |
| 3.00m | m | 12.80 | 192.00 | 0.00 | 28.80 | 220.80 |
| Extra for breaking up by hand | | | | | | |
| plain concrete 100mm thick | m2 | 0.90 | 13.50 | 0.00 | 2.03 | 15.53 |
| reinforced concrete 100mm thick | m2 | 1.00 | 15.00 | 0.00 | 2.25 | 17.25 |
| tarmacadam 75mm thick | m2 | 0.55 | 8.25 | 0.00 | 1.24 | 9.49 |
| hardcore 100mm thick | m2 | 0.40 | 6.00 | 0.00 | 0.90 | 6.90 |
| soft rock | m3 | 8.00 | 120.00 | 0.00 | 18.00 | 138.00 |
| hard rock | m3 | 10.00 | 150.00 | 0.00 | 22.50 | 172.50 |
| Sand bed in trench under pipe 100mm diameter, thickness | | | | | | |
| 100mm | m | 0.10 | 1.50 | 1.39 | 0.43 | 3.32 |
| 150mm | m | 0.12 | 1.80 | 2.04 | 0.58 | 4.42 |
| Sand bed in trench under pipe 150mm diameter, thickness | | | | | | |
| 100mm | m | 0.11 | 1.65 | 2.04 | 0.55 | 4.24 |
| 150mm | m | 0.13 | 1.95 | 3.11 | 0.76 | 5.82 |

| | Unit | Labour | Hours £ | Mat'ls £ | O & P £ | Total £ |
|---|---|---|---|---|---|---|
| Sand bed in trench under pipe 225mm diameter, thickness | | | | | | |
| 100mm | m | 0.14 | 2.10 | 2.14 | 0.64 | 4.88 |
| 150mm | m | 0.16 | 2.40 | 3.17 | 0.84 | 6.41 |
| Granular filling in bed in trench under pipe 100mm diameter, thickness | | | | | | |
| 100mm | m | 0.12 | 1.80 | 2.36 | 0.62 | 4.78 |
| 150mm | m | 0.14 | 2.10 | 3.49 | 0.84 | 6.43 |
| Granular filling in bed in trench under pipe 150mm diameter, thickness | | | | | | |
| 100mm | m | 0.13 | 1.95 | 2.45 | 0.66 | 5.06 |
| 150mm | m | 0.15 | 2.25 | 3.62 | 0.88 | 6.75 |
| Granular filling in bed in trench under pipe 225mm diameter, thickness | | | | | | |
| 100mm | m | 0.16 | 2.40 | 2.61 | 0.75 | 5.76 |
| 150mm | m | 0.18 | 2.70 | 3.77 | 0.97 | 7.44 |
| Concrete filling in bed in trench under pipe 100mm diameter, thickness | | | | | | |
| 100mm | m | 0.24 | 3.60 | 4.46 | 1.21 | 9.27 |
| 150mm | m | 0.28 | 4.20 | 6.64 | 1.63 | 12.47 |

| | Unit | Labour | Hours £ | Mat'ls £ | O & P £ | Total £ |
|---|---|---|---|---|---|---|
| Concrete filling in bed in trench under pipe 150mm diameter, thickness | | | | | | |
| 100mm | m | 0.26 | 3.90 | 5.24 | 1.37 | 10.51 |
| 150mm | m | 0.30 | 4.50 | 7.81 | 1.85 | 14.16 |
| Concrete filling in bed in trench under pipe 225mm diameter, thickness | | | | | | |
| 100mm | m | 0.32 | 4.80 | 5.92 | 1.61 | 12.33 |
| 150mm | m | 0.36 | 5.40 | 8.83 | 2.13 | 16.36 |
| Concrete in bed and haunching to pipe, 100mm diameter, bed thickness | | | | | | |
| 100mm | m | 0.48 | 7.20 | 5.92 | 1.97 | 15.09 |
| 150mm | m | 0.56 | 8.40 | 8.83 | 2.58 | 19.81 |
| Concrete in bed and haunching to pipe, 150mm diameter, bed thickness | | | | | | |
| 100mm | m | 0.52 | 7.80 | 7.57 | 2.31 | 17.68 |
| 150mm | m | 0.60 | 9.00 | 11.41 | 3.06 | 23.47 |
| Concrete in bed and haunching to pipe, 225mm diameter, bed thickness | | | | | | |
| 100mm | m | 0.60 | 9.00 | 10.19 | 2.88 | 22.07 |
| 150mm | m | 0.66 | 9.90 | 15.19 | 3.76 | 28.85 |

| | Unit | Labour | Hours £ | Mat'ls £ | O & P £ | Total £ |
|---|---|---|---|---|---|---|
| Granular filling in bed and surround to pipe, 100mm diameter, thickness | | | | | | |
| 100mm | m | 0.36 | 5.40 | 3.16 | 1.28 | 9.84 |
| 150mm | m | 0.42 | 6.30 | 4.69 | 1.65 | 12.64 |
| Granular filling in bed and surround to pipe, 150mm diameter, thickness | | | | | | |
| 100mm | m | 0.40 | 6.00 | 4.03 | 1.50 | 11.53 |
| 150mm | m | 0.46 | 6.90 | 6.01 | 1.94 | 14.85 |
| Granular filling in bed and surround to pipe, 225mm diameter, thickness | | | | | | |
| 100mm | m | 0.44 | 6.60 | 4.17 | 1.62 | 12.39 |
| 150mm | m | 0.50 | 7.50 | 6.21 | 2.06 | 15.77 |
| Concrete filling in bed and surround to pipe, 100mm diameter, thickness | | | | | | |
| 100mm | m | 0.72 | 10.80 | 15.33 | 3.92 | 30.05 |
| 150mm | m | 0.84 | 12.60 | 23.20 | 5.37 | 41.17 |
| Concrete filling in bed and surround to pipe, 150mm diameter, thickness | | | | | | |
| 100mm | m | 0.72 | 10.80 | 19.90 | 4.61 | 35.31 |
| 150mm | m | 0.84 | 12.60 | 29.79 | 6.36 | 48.75 |

| | Unit | Labour | Hours £ | Mat'ls £ | O & P £ | Total £ |
|---|---|---|---|---|---|---|
| Concrete filling in bed and surround to pipe, 225mm diameter, thickness | | | | | | |
| 100mm | m | 0.72 | 10.80 | 20.58 | 4.71 | 36.09 |
| 150mm | m | 0.84 | 12.60 | 32.44 | 6.76 | 51.80 |
| Hepworths' Supersleeve vitrified clay drain pipes, spigot and socket joints with sealing rings, 100mm diameter | | | | | | |
| laid in trenches | m | 0.36 | 5.40 | 6.99 | 1.86 | 14.25 |
| in lengths not exceeding 3m | m | 0.54 | 8.10 | 6.99 | 2.26 | 17.35 |
| bend | nr | 0.30 | 4.50 | 9.58 | 2.11 | 16.19 |
| rest bend | nr | 0.30 | 4.50 | 14.79 | 2.89 | 22.18 |
| single junction | nr | 0.30 | 4.50 | 13.96 | 2.77 | 21.23 |
| Hepworths' Supersleeve vitrified clay drain pipes, spigot and socket joints with sealing rings, 150mm diameter | | | | | | |
| laid in trenches | m | 0.60 | 9.00 | 13.24 | 3.34 | 25.58 |
| in lengths not exceeding 3m | m | 0.40 | 6.00 | 13.24 | 2.89 | 22.13 |
| bend | nr | 0.30 | 4.50 | 13.21 | 2.66 | 20.37 |
| rest bend | nr | 0.30 | 4.50 | 17.05 | 3.23 | 24.78 |
| single junction | nr | 0.30 | 4.50 | 17.76 | 3.34 | 25.60 |

| | Unit | Labour | Hours £ | Mat'ls £ | O & P £ | Total £ |
|---|---|---|---|---|---|---|
| Vitrified clay inlet gulley complete with grid and surrounded with concrete | nr | 1.50 | 22.50 | 43.96 | 9.97 | 76.43 |
| Vitrified clay back inlet gulley complete with grid and surrounded with concrete | nr | 1.50 | 22.50 | 62.74 | 12.79 | 98.03 |
| Vitrified clay paved area gulley complete with grid and surrounded with concrete | nr | 1.50 | 22.50 | 50.97 | 11.02 | 84.49 |

**Manholes**

| | Unit | Labour | Hours £ | Mat'ls £ | O & P £ | Total £ |
|---|---|---|---|---|---|---|
| Excavate by hand for manhole not exceeding | | | | | | |
| 1.0m deep | m3 | 4.00 | 60.00 | 0.00 | 9.00 | 69.00 |
| 1.5m deep | m3 | 4.50 | 67.50 | 0.00 | 10.13 | 77.63 |
| 2.0m deep | m3 | 5.60 | 84.00 | 0.00 | 12.60 | 96.60 |
| Load surplus excavated material into barrows wheel 50m and deposit into skip | m3 | 2.20 | 33.00 | 0.00 | 4.95 | 37.95 |
| Earthwork support not exceeding 2m between opposing faces, depth not exceeding | | | | | | |
| 1.0m deep | m2 | 0.35 | 5.25 | 2.17 | 1.11 | 8.53 |
| 2.0m deep | m2 | 0.40 | 6.00 | 2.49 | 1.27 | 9.76 |
| 4.0m deep | m2 | 0.45 | 6.75 | 2.80 | 1.43 | 10.98 |

| | Unit | Labour | Hours £ | Mat'ls £ | O & P £ | Total £ |
|---|---|---|---|---|---|---|
| Site-mixed concrete in base of manhole, thickness | | | | | | |
| 100-150mm | m3 | 2.00 | 30.00 | 96.21 | 18.93 | 145.14 |
| 150-300mm | m3 | 1.80 | 27.00 | 96.21 | 18.48 | 141.69 |
| Site-mixed concrete in benching to manhole, average thickness 225mm | m3 | 6.00 | 90.00 | 96.21 | 27.93 | 214.14 |
| Engineering bricks Class 'B' in cement mortar in walls of manhole | m2 | 4.80 | 72.00 | 62.40 | 20.16 | 154.56 |
| Extra for fair face and flush pointing | m2 | 0.25 | 3.75 | 0.00 | 0.56 | 4.31 |
| Build in ends of pipes to one brick wall and make good, pipe diameter | | | | | | |
| 100mm | nr | 0.15 | 2.25 | 0.00 | 0.34 | 2.59 |
| 150mm | nr | 0.19 | 2.84 | 0.00 | 0.43 | 3.26 |
| Galvanised step iron built into brickwork | nr | 0.20 | 3.00 | 5.94 | 1.34 | 10.28 |
| Cast iron manhole cover, frame, bedded in cement mortar | | | | | | |
| Grade A, light duty, size 600 × 450mm | nr | 1.90 | 28.50 | 77.36 | 15.88 | 121.74 |
| Grade B, medium duty, size 600 × 450mm | nr | 2.00 | 30.00 | 123.89 | 23.08 | 176.97 |

| | Unit | Labour | Hours £ | Mat'ls £ | O & P £ | Total £ |
|---|---|---|---|---|---|---|
| Best quality vitrified clay channels, bedded in cement mortar | | | | | | |
| half-section straight main channel | | | | | | |
| 100mm diameter × 300mm | nr | 0.75 | 11.25 | 3.27 | 2.18 | 16.70 |
| 100mm diameter × 600mm | nr | 0.85 | 12.75 | 5.84 | 2.79 | 21.38 |
| 100mm diameter × 1000mm | nr | 0.95 | 14.25 | 7.47 | 3.26 | 24.98 |
| 150mm diameter × 300mm | nr | 0.75 | 11.25 | 5.02 | 2.44 | 18.71 |
| 150mm diameter × 600mm | nr | 0.85 | 12.75 | 7.42 | 3.03 | 23.20 |
| 150mm diameter × 1000mm | nr | 0.95 | 14.25 | 10.28 | 3.68 | 28.21 |
| Half-section 90 degrees channel bend | | | | | | |
| 100mm | nr | 0.30 | 4.50 | 6.25 | 1.61 | 12.36 |
| 150mm | nr | 0.40 | 6.00 | 9.38 | 2.31 | 17.69 |
| Three-quarter section 90° channel bend | | | | | | |
| 100mm | nr | 1.20 | 18.00 | 13.08 | 4.66 | 35.74 |
| 150mm | nr | 1.30 | 19.50 | 23.27 | 6.42 | 49.19 |

| | Unit | Labour | Hours £ | Mat'ls £ | O & P £ | Total £ |
|---|---|---|---|---|---|---|
| **Land drainage** | | | | | | |
| Excavate trench 225mm wide by hand, remove excavated material, average depth | | | | | | |
| 500mm | m | 0.35 | 5.25 | 0.00 | 0.79 | 6.04 |
| 750mm | m | 0.50 | 7.50 | 0.00 | 1.13 | 8.63 |
| 1000mm | m | 0.75 | 11.25 | 0.00 | 1.69 | 12.94 |
| Excavate trench 300mm wide by hand, remove excavated material, average depth | | | | | | |
| 500mm | m | 0.45 | 6.75 | 0.00 | 1.01 | 7.76 |
| 750mm | m | 0.70 | 10.50 | 0.00 | 1.58 | 12.08 |
| 1000mm | m | 0.90 | 13.50 | 0.00 | 2.03 | 15.53 |
| Agricultural clay field drain pipe, 300mm long, laid butt jointed in trench, diameter | | | | | | |
| 75mm | m | 0.11 | 1.65 | 2.99 | 0.70 | 5.34 |
| 100mm | m | 0.12 | 1.80 | 5.02 | 1.02 | 7.84 |
| 150mm | m | 0.14 | 2.10 | 9.57 | 1.75 | 13.42 |
| Single junction | | | | | | |
| 75mm | nr | 0.06 | 0.90 | 10.53 | 1.71 | 13.14 |
| 100mm | nr | 0.07 | 1.05 | 13.69 | 2.21 | 16.95 |
| 150mm | nr | 0.08 | 1.20 | 16.88 | 2.71 | 20.79 |

| | Unit | Labour | Hours £ | Mat'ls £ | O & P £ | Total £ |
|---|---|---|---|---|---|---|
| Flexible plastic perforated field drain pipe, diameter | | | | | | |
| 60mm | m | 0.02 | 0.30 | 0.36 | 0.10 | 0.76 |
| 80mm | m | 0.02 | 0.30 | 0.47 | 0.12 | 0.89 |
| 100mm | m | 0.03 | 0.45 | 0.74 | 0.18 | 1.37 |
| Single junction | | | | | | |
| 60 × 60mm | nr | 0.06 | 0.90 | 1.68 | 0.39 | 2.97 |
| 60 × 80mm | nr | 0.06 | 0.90 | 1.82 | 0.41 | 3.13 |
| 60 × 100mm | nr | 0.06 | 0.90 | 1.86 | 0.41 | 3.17 |
| 80 × 80mm | nr | 0.06 | 0.90 | 1.80 | 0.41 | 3.11 |
| 80 × 100mm | nr | 0.06 | 0.90 | 1.93 | 0.42 | 3.25 |
| Backfilling field trench with gravel, blinded with sand and replace with turf laid aside | | | | | | |
| trench width 225mm | | | | | | |
| depth 500mm | m | 0.03 | 0.45 | 2.28 | 0.41 | 3.14 |
| depth 750mm | m | 0.03 | 0.45 | 3.33 | 0.57 | 4.35 |
| depth 1000mm | m | 0.03 | 0.45 | 4.35 | 0.72 | 5.52 |
| trench width 300mm | | | | | | |
| depth 500mm | m | 0.03 | 0.45 | 3.09 | 0.53 | 4.07 |
| depth 750mm | m | 0.03 | 0.45 | 4.36 | 0.72 | 5.53 |
| depth 1000mm | m | 0.03 | 0.45 | 5.71 | 0.92 | 7.08 |

# Part Three

## GROUNDWORKS

Roads and sewers

House foundations

| | Unit | Plant £ | Mat'ls £ | O & P £ | Total £ |
|---|---|---|---|---|---|

This Groundworks section covers work in small housing estates including roads, sewers and house foundations.

**ROADS AND SEWERS**

Where applicable the plant column includes the cost of the operator.

The following rates are based on excavating in firm ground. The following adjustments should be made for other conditions:

stiff clay + 50%
soft chalk + 100%

| | Unit | Plant £ | Mat'ls £ | O & P £ | Total £ |
|---|---|---|---|---|---|
| Remove undergrowth and site vegetation | m2 | 0.17 | - | 0.03 | 0.20 |
| Cut down trees, grub up roots and remove | | | | | |
| girth, 600–1500mm | nr | 200.00 | - | 30.00 | 230.00 |
| girth, 1500–3000mm | nr | 527.60 | - | 79.14 | 606.74 |

| | Unit | Plant £ | Mat'ls £ | O & P £ | Total £ |
|---|---|---|---|---|---|
| Cut down hedge, grub up roots and remove | | | | | |
| height, 1500mm | m | 17.36 | - | 2.60 | 19.96 |
| height, 3000mm | m | 30.40 | - | 4.56 | 34.96 |
| Excavate topsoil or turf and lay aside for re-use | | | | | |
| 150mm thick | m2 | 0.33 | - | 0.05 | 0.38 |
| 200mm thick | m2 | 0.44 | - | 0.07 | 0.51 |
| 250mm thick | m2 | 0.54 | - | 0.08 | 0.62 |
| Excavate to reduce levels depth not exceeding | | | | | |
| 250mm thick | m3 | 1.62 | - | 0.24 | 1.86 |
| 500mm thick | m3 | 1.52 | - | 0.23 | 1.75 |
| Extra for excavating through | | | | | |
| rock | m3 | 71.65 | - | 10.75 | 82.40 |
| concrete | m3 | 60.80 | - | 9.12 | 69.92 |
| brickwork | m3 | 47.76 | - | 7.16 | 54.92 |

| | Unit | Labour | Hours £ | Mat'ls £ | O & P £ | Total £ |
|---|---|---|---|---|---|---|
| Excavate trench for kerb foundation | | | | | | |
| 200 x 75mm thick | m | 0.10 | 1.50 | - | 0.23 | 1.73 |
| 250 x 100mm thick | m | 0.12 | 1.80 | - | 0.27 | 2.07 |
| 300 x 100mm thick | m | 0.14 | 2.10 | - | 0.32 | 2.42 |
| 450 x 500mm thick | m | 0.20 | 3.00 | - | 0.45 | 3.45 |

| | Unit | Plant £ | Mat'ls £ | O & P £ | Total £ |
|---|---|---|---|---|---|
| **Disposal by machine** | | | | | |
| Load surplus excavated material into lorries and cart away to tip | | | | | |
| distance, 10km | m3 | 17.96 | - | 2.69 | 20.65 |
| distance, 15km | m3 | 19.54 | - | 2.93 | 22.47 |
| **Filling by machine** | | | | | |
| Surplus excavated material deposited and compacted in layers | | | | | |
| over 250mm thick | m3 | 7.84 | - | 1.18 | 9.02 |
| 100mm thick | m2 | 1.20 | - | 0.18 | 1.38 |
| 150mm thick | m2 | 1.35 | - | 0.20 | 1.55 |
| 200mm thick | m2 | 1.50 | - | 0.23 | 1.73 |
| Imported sand deposited and compacted in layers | | | | | |
| over 250mm thick | m3 | 7.84 | 33.91 | 6.26 | 48.01 |
| 100mm thick | m2 | 1.20 | 3.39 | 0.69 | 5.28 |
| 150mm thick | m2 | 1.35 | 5.08 | 0.96 | 7.39 |
| 200mm thick | m2 | 1.50 | 7.78 | 1.39 | 10.67 |
| Imported hardcore deposited and compacted in layers | | | | | |
| over 250mm thick | m3 | 7.84 | 20.02 | 4.18 | 32.04 |
| 100mm thick | m2 | 1.20 | 2.02 | 0.48 | 3.70 |
| 150mm thick | m2 | 1.35 | 3.03 | 0.66 | 5.04 |
| 200mm thick | m2 | 1.50 | 4.04 | 0.83 | 6.37 |

| | Unit | Plant £ | Mat'ls £ | O & P £ | Total £ |
|---|---|---|---|---|---|
| Imported topsoil deposited and compacted in layers | | | | | |
| over 250mm thick | m3 | 7.84 | 15.20 | 3.46 | 26.50 |
| 100mm thick | m2 | 1.20 | 1.52 | 0.41 | 3.13 |
| 150mm thick | m2 | 1.35 | 2.28 | 0.54 | 4.17 |
| 200mm thick | m2 | 1.50 | 3.04 | 0.68 | 5.22 |

| | Unit | Labour | Hours £ | Mat'ls £ | O & P £ | Total £ |
|---|---|---|---|---|---|---|
| Precast concrete kerbs, splayed or battered, laid straight or curved to radius exceeding 12m | | | | | | |
| 125 x 150mm | m | 0.40 | 6.00 | 10.12 | 2.42 | 18.54 |
| 125 x 150mm | m | 0.45 | 6.75 | 11.89 | 2.80 | 21.44 |
| 150 x 305mm | m | 0.50 | 7.50 | 13.25 | 3.11 | 23.86 |
| Precast concrete kerbs, splayed or battered, laid straight or curved to radius not exceeding 12m | | | | | | |
| 125 x 150mm | m | 0.40 | 6.00 | 10.12 | 2.42 | 18.54 |
| 125 x 150mm | m | 0.45 | 6.75 | 11.89 | 2.80 | 21.44 |
| 150 x 305mm | m | 0.50 | 7.50 | 13.25 | 3.11 | 23.86 |

| | Unit | Plant £ | Mat'ls £ | O & P £ | Total £ |
|---|---|---|---|---|---|
| Granular material in sub-base to road, depth | | | | | |
| 75mm | m2 | 1.20 | 1.07 | 0.34 | 2.61 |
| 100mm | m2 | 1.40 | 1.42 | 0.42 | 3.24 |
| 150mm | m2 | 1.80 | 2.13 | 0.59 | 4.52 |
| 200mm | m2 | 2.40 | 2.84 | 0.79 | 6.03 |
| 250mm | m2 | 2.60 | 3.56 | 0.92 | 7.08 |
| Lean concrete in road base, depth | | | | | |
| 100mm | m2 | 1.40 | 7.55 | 1.34 | 10.29 |
| 150mm | m2 | 1.80 | 11.33 | 1.97 | 15.10 |
| 200mm | m2 | 2.40 | 15.10 | 2.63 | 20.13 |
| 250mm | m2 | 2.60 | 18.88 | 3.22 | 24.70 |
| Dense bitumen macadam road base, depth | | | | | |
| 75mm | m2 | 4.10 | 4.74 | 1.33 | 10.17 |
| 100mm | m2 | 6.50 | 6.32 | 1.92 | 14.74 |
| 150mm | m2 | 1.80 | 2.13 | 0.59 | 4.52 |
| 200mm | m2 | 2.40 | 2.84 | 0.79 | 6.03 |
| Rolled asphalt base course, depth | | | | | |
| 50mm | m2 | 5.61 | 3.55 | 1.37 | 10.53 |
| 75mm | m2 | 8.36 | 4.92 | 1.99 | 15.27 |

| | Unit | Plant £ | Mat'ls £ | O & P £ | Total £ |
|---|---|---|---|---|---|
| Dense tar surfacing werаring course, depth | | | | | |
| 25mm | m2 | 2.94 | 2.32 | 0.79 | 6.05 |
| 50mm | m2 | 4.22 | 3.88 | 1.22 | 9.32 |
| Cold asphalt wearing course, depth | | | | | |
| 15mm | m2 | 4.21 | 1.30 | 0.83 | 6.34 |
| 25mm | m2 | 5.22 | 2.04 | 1.09 | 8.35 |
| Rolled asphalt wearing course, depth | | | | | |
| 30mm | m2 | 4.21 | 2.93 | 1.07 | 8.21 |
| 50mm | m2 | 5.22 | 4.70 | 1.49 | 11.41 |
| Concrete, designed mix grade 30, 20 mm aggregate, depth | | | | | |
| 100mm | m2 | 3.84 | 8.20 | 1.81 | 13.85 |
| 150mm | m2 | 7.60 | 12.30 | 2.99 | 22.89 |
| 200mm | m2 | 9.77 | 16.40 | 3.93 | 30.10 |
| 250mm | m2 | 10.44 | 20.48 | 4.64 | 35.56 |
| 300mm | m2 | 11.63 | 24.58 | 5.43 | 41.64 |
| 400mm | m2 | 13.28 | 32.77 | 6.91 | 52.96 |
| Steel fabric reinforcement to BS 4483 | | | | | |
| ref A142, weight 2.26kg/m2 | m2 | 2.90 | 1.36 | 0.64 | 4.90 |
| ref A193, weight 3.02kg/m2 | m2 | 4.88 | 1.78 | 1.00 | 7.66 |
| ref C503, weight 4.34kg/m2 | m2 | 6.70 | 2.72 | 1.41 | 10.83 |
| ref C636, weight 5.55kg/m2 | m2 | 7.65 | 3.44 | 1.66 | 12.75 |

| | Unit | Labour | Hours £ | Mat'ls £ | O & P £ | Total £ |
|---|---|---|---|---|---|---|
| Waterproof membrane below concrete slab | | | | | | |
| plastic sheeting, 250 micron | m2 | 0.10 | 1.50 | 0.50 | 0.30 | 2.30 |
| plastic sheeting, 500 micron | m2 | 0.10 | 1.50 | 0.50 | 0.30 | 2.30 |
| Longitudinal joints, depth | | | | | | |
| 150mm | m | 0.16 | 2.40 | 27.10 | 4.43 | 33.93 |
| 200mm | m | 0.16 | 2.40 | 29.54 | 4.79 | 36.73 |
| 250mm | m | 0.16 | 2.40 | 32.88 | 5.29 | 40.57 |
| 300mm | m | 0.16 | 2.40 | 32.47 | 5.23 | 40.10 |
| Expansion joints, depth | | | | | | |
| 150mm | m | 0.16 | 2.40 | 33.56 | 5.39 | 41.35 |
| 200mm | m | 0.16 | 2.40 | 36.68 | 5.86 | 44.94 |
| 250mm | m | 0.16 | 2.40 | 39.00 | 6.21 | 47.61 |
| 300mm | m | 0.16 | 2.40 | 42.37 | 6.72 | 51.49 |
| Contraction joints, depth | | | | | | |
| 150mm | m | 0.16 | 2.40 | 17.55 | 2.99 | 22.94 |
| 200mm | m | 0.16 | 2.40 | 19.64 | 3.31 | 25.35 |
| 250mm | m | 0.16 | 2.40 | 22.57 | 3.75 | 28.72 |
| 300mm | m | 0.16 | 2.40 | 25.67 | 4.21 | 32.28 |
| Construction joints, depth | | | | | | |
| 150mm | m | 0.16 | 2.40 | 16.22 | 2.79 | 21.41 |
| 200mm | m | 0.16 | 2.40 | 17.88 | 3.04 | 23.32 |
| 250mm | m | 0.16 | 2.40 | 20.22 | 3.39 | 26.01 |
| 300mm | m | 0.16 | 2.40 | 20.88 | 3.49 | 26.77 |

| | Unit | Labour | Hours £ | Mat'ls £ | O & P £ | Total £ |
|---|---|---|---|---|---|---|
| Vitrified clay road gulley surrounded with concrete, two courses of brickwork, and cast iron grating | | | | | | |
| 450 x 750mm | m | 1.20 | 18.00 | 128.36 | 21.95 | 168.31 |
| 450 x 900mm | m | 1.20 | 18.00 | 132.55 | 22.58 | 173.13 |
| 450 x 1050mm | m | 1.20 | 18.00 | 138.97 | 23.55 | 180.52 |
| 450 x 915mm | m | 1.20 | 18.00 | 370.55 | 58.28 | 446.83 |
| Concrete road gulley surrounded with concrete, two courses of brickwork, and cast iron grating | | | | | | |
| 450 x 750mm | m | 1.20 | 18.00 | 136.46 | 23.17 | 177.63 |
| 450 x 900mm | m | 1.20 | 18.00 | 128.69 | 22.00 | 168.69 |
| 450 x 1050mm | m | 1.20 | 18.00 | 133.53 | 22.73 | 174.26 |
| 450 x 915mm | m | 1.20 | 18.00 | 145.65 | 24.55 | 188.20 |

| | Unit | Plant £ | Mat'ls £ | O & P £ | Total £ |
|---|---|---|---|---|---|
| Concrete pipes, Class L, with flexible joints, nominal bore | | | | | |
| 300mm, depth | | | | | |
| 1.5–2m | m | 34.55 | 13.60 | 7.22 | 55.37 |
| 2–2.5m | m | 36.56 | 13.60 | 7.52 | 57.68 |
| 2.5–3m | m | 38.14 | 13.60 | 7.76 | 59.50 |

| | Unit | Plant £ | Mat'ls £ | O & P £ | Total £ |
|---|---|---|---|---|---|
| 450mm, depth | | | | | |
| 1.5–2m | m | 38.14 | 20.34 | 8.77 | 67.25 |
| 2–2.5m | m | 40.22 | 20.34 | 9.08 | 69.64 |
| 2.5–3m | m | 44.85 | 20.34 | 9.78 | 74.97 |
| 600mm, depth | | | | | |
| 1.5–2m | m | 58.76 | 32.20 | 13.64 | 104.60 |
| 2–2.5m | m | 66.00 | 32.20 | 14.73 | 112.93 |
| 2.5–3m | m | 73.20 | 32.20 | 15.81 | 121.21 |
| Concrete pipe fittings, nominal bore | | | | | |
| 300mm | | | | | |
| bends, 45 degrees | nr | 11.58 | 114.51 | 18.91 | 145.00 |
| junctions | nr | 14.38 | 88.49 | 15.43 | 118.30 |
| 450mm | | | | | |
| bends, 45 degrees | nr | 14.46 | 168.55 | 27.45 | 210.46 |
| junctions | nr | 20.32 | 114.56 | 20.23 | 155.11 |
| 600mm | | | | | |
| bends, 45 degrees | nr | 18.37 | 384.25 | 60.39 | 463.01 |
| junctions | nr | 28.57 | 114.56 | 21.47 | 164.60 |

| | Unit | Labour | Hours £ | Mat'ls £ | O & P £ | Total £ |
|---|---|---|---|---|---|---|
| **HOUSE FOUNDATIONS** | | | | | | |
| **Hand excavation** | | | | | | |
| Excavate topsoil | | | | | | |
| 150mm thick | m2 | 0.35 | 5.25 | - | 0.79 | 6.04 |
| 200mm thick | m2 | 0.45 | 6.75 | - | 1.01 | 7.76 |
| 250mm thick | m2 | 0.60 | 9.00 | - | 1.35 | 10.35 |
| Excavate to reduce levels, levels, depth not exceeding | | | | | | |
| 0.75m | m3 | 2.20 | 33.00 | - | 4.95 | 37.95 |
| 1.00m | m3 | 2.40 | 36.00 | - | 5.40 | 41.40 |
| 1.50m | m3 | 2.80 | 42.00 | - | 6.30 | 48.30 |
| 2.00m | m3 | 3.20 | 48.00 | - | 7.20 | 55.20 |
| Excavate trenches 450mm wide,depth not exceeding | | | | | | |
| 0.75m | m3 | 2.50 | 37.50 | - | 5.63 | 43.13 |
| 1.00m | m3 | 2.70 | 40.50 | - | 6.08 | 46.58 |
| 1.50m | m3 | 3.00 | 45.00 | - | 6.75 | 51.75 |
| 2.00m | m3 | 3.40 | 51.00 | - | 7.65 | 58.65 |
| Extra for excavating through | | | | | | |
| rock | m3 | 10.00 | 150.00 | - | 22.50 | 172.50 |
| concrete | m3 | 8.00 | 120.00 | - | 18.00 | 138.00 |
| reinforced concrete | m3 | 9.00 | 135.00 | - | 20.25 | 155.25 |
| brickwork | m3 | 6.00 | 90.00 | - | 13.50 | 103.50 |

| | Unit | Plant £ | Mat'ls £ | O & P £ | Total £ |
|---|---|---|---|---|---|
| **Excavate by machine** | | | | | |
| Excavate topsoil | | | | | |
| 150mm thick | m2 | 1.90 | - | 0.29 | 2.19 |
| 200mm thick | m2 | 2.55 | - | 0.38 | 2.93 |
| 250mm thick | m2 | 3.34 | - | 0.50 | 3.84 |
| Excavate to reduce levels, levels, depth not exceeding | | | | | |
| 0.75m | m3 | 2.70 | - | 0.41 | 3.11 |
| 1.00m | m3 | 2.72 | - | 0.41 | 3.13 |
| 1.50m | m3 | 3.22 | - | 0.48 | 3.70 |
| 2.00m | m3 | 3.64 | - | 0.55 | 4.19 |
| Excavate trenches 450mm wide, depth not exceeding | | | | | |
| 0.75m | m3 | 8.25 | - | 1.24 | 9.49 |
| 1.00m | m3 | 8.86 | - | 1.33 | 10.19 |
| 1.50m | m3 | 9.67 | - | 1.45 | 11.12 |
| 2.00m | m3 | 10.55 | - | 1.58 | 12.13 |
| Extra for excavating through | | | | | |
| rock | m3 | 78.00 | - | 11.70 | 89.70 |
| concrete | m3 | 58.00 | - | 8.70 | 66.70 |
| reinforced concrete | m3 | 65.21 | - | 9.78 | 74.99 |
| brickwork | m3 | 52.32 | - | 7.85 | 60.17 |

| | Unit | Labour | Hours £ | Mat'ls £ | O & P £ | Total £ |
|---|---|---|---|---|---|---|
| **Earthwork support** | | | | | | |
| Earthwork support not exceeding 2m between opposite faces, depth depth not exceeding 1m | | | | | | |
| firm ground | m2 | 0.15 | 2.25 | 1.57 | 0.57 | 4.39 |
| loose ground | m2 | 0.85 | 12.75 | 3.02 | 2.37 | 18.14 |
| sand | m2 | 1.10 | 16.50 | 3.58 | 3.01 | 23.09 |
| Earthwork support not exceeding 2m between opposite faces, depth depth not exceeding 2m | | | | | | |
| firm ground | m2 | 0.18 | 2.70 | 1.57 | 0.64 | 4.91 |
| loose ground | m2 | 0.90 | 13.50 | 3.02 | 2.48 | 19.00 |
| sand | m2 | 1.20 | 18.00 | 3.58 | 3.24 | 24.82 |
| Earthwork support not exceeding 2m between opposite faces, depth depth not exceeding 4m | | | | | | |
| firm ground | m2 | 0.22 | 3.30 | 1.57 | 0.73 | 5.60 |
| loose ground | m2 | 0.95 | 14.25 | 3.02 | 2.59 | 19.86 |
| sand | m2 | 1.30 | 19.50 | 3.58 | 3.46 | 26.54 |

| | Unit | Labour | Hours £ | Mat'ls £ | O & P £ | Total £ |
|---|---|---|---|---|---|---|
| **Disposal by hand** | | | | | | |
| Load surplus excavated material into barrows, wheel and deposit in temporary spoil heaps, average distance | | | | | | |
| 15m | m3 | 1.25 | 18.75 | - | 2.81 | 21.56 |
| 25m | m3 | 1.45 | 21.75 | - | 3.26 | 25.01 |
| 50m | m3 | 1.70 | 25.50 | - | 3.83 | 29.33 |
| Load surplus excavated material into barrows, wheel and spread over site, average distance | | | | | | |
| 15m | m3 | 1.65 | 24.75 | - | 3.71 | 28.46 |
| 25m | m3 | 1.85 | 27.75 | - | 4.16 | 31.91 |
| 50m | m3 | 2.00 | 30.00 | - | 4.50 | 34.50 |
| Load surplus excavated material into barrows, wheel and deposit in skips or lorries, average distance | | | | | | |
| 15m | m3 | 1.20 | 18.00 | - | 2.70 | 20.70 |
| 25m | m3 | 1.40 | 21.00 | - | 3.15 | 24.15 |
| 50m | m3 | 1.65 | 24.75 | - | 3.71 | 28.46 |

| | Unit | Plant £ | Mat'ls £ | O & P £ | Total £ |
|---|---|---|---|---|---|
| **Disposal by machine** | | | | | |
| Load and deposit in temporary spoil heaps, average distance | | | | | |
| 15m | m3 | 2.45 | - | 0.37 | 2.82 |
| 25m | m3 | 2.98 | - | 0.45 | 3.43 |
| 50m | m3 | 4.92 | - | 0.74 | 5.66 |
| Load and spread over site, average distance | | | | | |
| 15m | m3 | 3.10 | - | 0.47 | 3.57 |
| 25m | m3 | 3.84 | - | 0.58 | 4.42 |
| 50m | m3 | 4.38 | - | 0.66 | 5.04 |
| Load and deposit in skips or lorries, average distance | | | | | |
| 15m | m3 | 2.16 | - | 0.32 | 2.48 |
| 25m | m3 | 2.65 | - | 0.40 | 3.05 |
| 50m | m3 | 4.44 | - | 0.67 | 5.11 |

| | Unit | Labour | Hours £ | Mat'ls £ | O & P £ | Total £ |
|---|---|---|---|---|---|---|
| **Filling by hand** | | | | | | |
| Surplus excavated material deposited and compacting in layers | | | | | | |
| over 250mm thick | m3 | 0.80 | 12.00 | - | 1.80 | 13.80 |
| 100mm thick | m2 | 0.10 | 1.50 | - | 0.23 | 1.73 |
| 150mm thick | m2 | 0.15 | 2.25 | - | 0.34 | 2.59 |
| 200mm thick | m2 | 0.20 | 3.00 | - | 0.45 | 3.45 |
| Imported sand deposited and compacting in layers | | | | | | |
| over 250mm thick | m3 | 0.80 | 12.00 | 24.56 | 5.48 | 42.04 |
| 100mm thick | m2 | 0.10 | 1.50 | 2.45 | 0.59 | 4.54 |
| 150mm thick | m2 | 0.15 | 2.25 | 3.68 | 0.89 | 6.82 |
| 200mm thick | m2 | 0.20 | 3.00 | 4.90 | 1.19 | 9.09 |
| Imported hardcore deposited and compacting in layers | | | | | | |
| over 250mm thick | m3 | 0.80 | 12.00 | 22.46 | 5.17 | 39.63 |
| 100mm thick | m2 | 0.10 | 1.50 | 2.24 | 0.56 | 4.30 |
| 150mm thick | m2 | 0.15 | 2.25 | 3.36 | 0.84 | 6.45 |
| 200mm thick | m2 | 0.20 | 3.00 | 4.48 | 1.12 | 8.60 |

| | Unit | Plant £ | Mat'ls £ | O & P £ | Total £ |
|---|---|---|---|---|---|
| **Filling by machine** | | | | | |
| Surplus excavated material deposited and compacting in layers | | | | | |
| over 250mm thick | m3 | 5.08 | - | 0.76 | 5.84 |
| 100mm thick | m2 | 0.78 | - | 0.12 | 0.90 |
| 150mm thick | m2 | 1.03 | - | 0.15 | 1.18 |
| 200mm thick | m2 | 1.22 | - | 0.18 | 1.40 |
| Imported sand deposited and compacting in layers | | | | | |
| over 250mm thick | m3 | 5.08 | 32.44 | 5.63 | 43.15 |
| 100mm thick | m2 | 0.78 | 3.24 | 0.60 | 4.62 |
| 150mm thick | m2 | 1.03 | 4.86 | 0.88 | 6.77 |
| 200mm thick | m2 | 1.22 | 6.48 | 1.16 | 8.86 |
| Imported hardcore deposited and compacting in layers | | | | | |
| over 250mm thick | m3 | 5.08 | 21.24 | 3.95 | 30.27 |
| 100mm thick | m2 | 0.78 | 2.12 | 0.44 | 3.34 |
| 150mm thick | m2 | 1.03 | 3.18 | 0.63 | 4.84 |
| 200mm thick | m2 | 1.22 | 4.24 | 0.82 | 6.28 |
| **Surface treatments** | | | | | |
| Level and compact bottom of excavation with vibrating roller | m2 | 1.28 | - | 0.19 | 1.47 |
| Blind filling surfaces with sand 50mm thick | m2 | 1.24 | 1.16 | 0.36 | 2.76 |

| | Unit | Labour | Hours £ | Mat'ls £ | O & P £ | Total £ |
|---|---|---|---|---|---|---|
| **Concrete work** | | | | | | |
| Ready mix concrete 1:3:6 in foundation trenches, thickness | | | | | | |
| 150 to 300mm | m3 | 0.90 | 13.50 | 81.25 | 14.21 | 108.96 |
| 150 to 300mm | m3 | 0.70 | 10.50 | 81.25 | 13.76 | 105.51 |
| 300 to 450mm | m3 | 0.50 | 7.50 | 81.25 | 13.31 | 102.06 |
| Ready mix concrete 1:2:4 in ground floor slabs, thickness | | | | | | |
| 100 to 300mm | m3 | 1.10 | 16.50 | 83.24 | 14.96 | 114.70 |
| 300 to 450mm | m3 | 1.00 | 15.00 | 83.24 | 14.74 | 112.98 |
| **Brickwork** | | | | | | |
| Common bricks basic price £140 per 1,000 in cement mortar | | | | | | |
| half brick thick | m2 | 0.65 | 31.27 | 11.03 | 6.34 | 48.64 |
| one brick thick | m2 | 1.10 | 52.91 | 22.90 | 11.37 | 87.18 |
| Facing bricks basic price £400 per 1,000 in cement mortar | | | | | | |
| half brick thick | m2 | 0.85 | 40.89 | 27.41 | 10.24 | 78.54 |
| one brick thick | m2 | 1.10 | 52.91 | 55.41 | 16.25 | 124.57 |

| | Unit | Labour | Hours £ | Mat'ls £ | O & P £ | Total £ |
|---|---|---|---|---|---|---|
| **Blockwork** | | | | | | |
| Dense concrete blockwork in cement mortar | | | | | | |
| 100mm thick | m2 | 0.40 | 19.24 | 9.08 | 4.25 | 32.57 |
| 140mm thick | m2 | 0.46 | 22.13 | 12.24 | 5.15 | 39.52 |

| | Unit | Quantity | Rate | Total £ |
|---|---|---|---|---|
| The following are typical costs for the construction of the the foundations for three different sized houses. The figures have been rounded off. | | | | |
| **Ground floor area 63m2** | | | | |
| Excavate topsoil 150mm deep by machine | m2 | 63 | 2.19 | 138 |
| Excavate to reduce levels 0.75m deep | m3 | 47 | 3.11 | 146 |
| Excavate trench 450mm wide not exceeding 1m deep | m3 | 18 | 10.19 | 183 |
| Carried forward | | | | 468 |

| | Unit | Quantity | Rate | Total £ |
|---|---|---|---|---|
| Brought forward | | | | 468 |
| Earthwork support not exceeding 2m between faces not exceeding 1m deep | m3 | 80 | 4.91 | 393 |
| Load and deposit surplus excavated material in lorries, average distance 15m | m3 | 47 | 2.48 | 117 |
| Level and compact bottom of trench | m2 | 18 | 1.47 | 26 |
| Ready mix concrete 1:3:6 in foundations 150 to 300mm deep | m3 | 4 | 108.96 | 436 |
| Ready mix concrete 1:2:4 in floor slab 150 to 300mm deep | m3 | 9 | 114.70 | 1,032 |
| Half brick wall in common bricks £140 per thousand in leaf of cavity wall in cement mortar | m2 | 24 | 48.64 | 1,167 |
| Half brick wall in facing bricks £400 per thousand in leaf of cavity wall in cement mortar | m2 | 6 | 78.54 | 471 |
| Carried forward | | | | 4,110 |

| | Unit | Quantity | Rate | Total £ |
|---|---|---|---|---|
| Brought forward | | | | 4,110 |
| Dense concrete blockwork in cavity wall in cement mortar | m2 | 30 | 32.57 | 977 |
| **Total** | | | | 5,087 |
| **Ground floor area 102m2** | | | | |
| Excavate topsoil 150mm deep by machine | m2 | 102 | 2.19 | 223 |
| Excavate to reduce levels 0.75m deep | m3 | 76 | 3.11 | 236 |
| Excavate trench 450mm wide not exceeding 1m deep | m3 | 26 | 10.19 | 265 |
| Earthwork support not exceeding 2m between faces not exceeding 1m deep | m3 | 114 | 4.91 | 560 |
| Load and deposit surplus excavated material in lorries, average distance 15m | m3 | 76 | 2.48 | 188 |
| Level and compact bottom of trench | m2 | 18 | 1.47 | 26 |
| Carried forward | | | | 1499 |

| | Unit | Quantity | Rate | Total £ |
|---|---|---|---|---|
| | Brought forward | | | 1,499 |
| Ready mix concrete 1:3:6 in foundations 150 to 300mm deep | m3 | 6 | 108.96 | 654 |
| Ready mix concrete 1:2:4 in floor slab 150 to 300mm deep | m3 | 15 | 114.70 | 1,721 |
| Half brick wall in common bricks £140 per thousand in leaf of cavity wall in cement mortar | m2 | 34 | 48.64 | 1,654 |
| Half brick wall in facing bricks £400 per thousand in leaf of cavity wall in cement mortar | m2 | 9 | 78.54 | 707 |
| Dense concrete blockwork in cavity wall in cement mortar | m2 | 42 | 32.57 | 1,368 |
| | **Total** | | | 7,602 |
| **Ground floor area 250m2** | | | | |
| Excavate topsoil 150mm deep by machine | m2 | 250 | 2.19 | 548 |
| Excavate to reduce levels 0.75m deep | m3 | 188 | 3.11 | 585 |
| | Carried forward | | | 1132 |

| | Unit | Quantity | Rate | Total £ |
|---|---|---|---|---|
| Brought forward | | | | 1,132 |
| Excavate trench 450mm wide not exceeding 1m deep | m3 | 36 | 10.19 | 367 |
| Earthwork support not exceeding 2m between faces not exceeding 1m deep | m3 | 158 | 4.91 | 776 |
| Load and deposit surplus excavated material in lorries, average distance 15m | m3 | 188 | 2.48 | 466 |
| Level and compact bottom of trench | m2 | 36 | 1.47 | 53 |
| Ready mix concrete 1:3:6 in foundations 150 to 300mm deep | m3 | 8 | 108.96 | 872 |
| Ready mix concrete 1:2:4 in floor slab 150 to 300mm deep | m3 | 38 | 114.70 | 4,359 |
| Half brick wall in common bricks £140 per thousand in leaf of cavity wall in cement mortar | m2 | 47 | 48.64 | 2,286 |
| Carried forward | | | | 10,310 |

| | Unit | Quantity | Rate | Total £ |
|---|---|---|---|---|
| Brought forward | | | | 10,310 |
| Half brick wall in facing bricks £400 per thousand in leaf of cavity wall in cement mortar | m2 | 12 | 78.54 | 942 |
| Dense concrete blockwork in cavity wall in cement mortar | m2 | 38 | 32.57 | 1,238 |
| | **Total** | | | 12,490 |

# Part Four

## APPROXIMATE ESTIMATING

Brick walling

Masonry

Kerbs and edgings

Beds and pavings

Drainage

| | Unit | Rate £ |
|---|---|---|
| **BRICK WALLING** | | |
| Excavate trench, lay concrete foundation, build wall 1m high in common bricks (basic price £140 per 1000) in cement mortar (1:3), stretcher bond | | |
| half brick thick | m | 82.00 |
| half brick thick, curved | m | 96.00 |
| one brick thick | m | 120.00 |
| one thick, curved | m | 134.00 |
| one and a half brick thick | m | 150.00 |
| two brick thick | m | 175.00 |
| Excavate trench, lay concrete foundation, build wall 1m high in common bricks (basic price £140 per 1000) in cement mortar (1:3), stretcher bond, facework one side | | |
| half brick thick | m | 84.00 |
| half brick thick, curved | m | 98.00 |
| one brick thick | m | 122.00 |
| one thick, curved | m | 136.00 |
| one and a half brick thick | m | 152.00 |
| two brick thick | m | 177.00 |
| Excavate trench, lay concrete foundation, build wall 1m high in common bricks (basic price £140 per 1000) in cement mortar (1:3), stretcher bond, facework both sides | | |
| half brick thick | m | 86.00 |
| half brick thick, curved | m | 100.00 |

| | Unit | Rate £ |
|---|---|---|
| **Brick walling (cont'd)** | | |
| one brick thick | m | 124.00 |
| one thick, curved | m | 138.00 |
| one and a half brick thick | m | 154.00 |
| two brick thick | m | 180.00 |
| Excavate trench, lay concrete foundation, build wall 1m high in common bricks (basic price £200 per 1000) in cement mortar (1:3), stretcher bond | | |
| half brick thick | m | 88.00 |
| half brick thick, curved | m | 102.00 |
| one brick thick | m | 130.00 |
| one thick, curved | m | 144.00 |
| one and a half brick thick | m | 168.00 |
| two brick thick | m | 204.00 |
| Excavate trench, lay concrete foundation, build wall 1m high in common bricks (basic price £200 per 1000) in cement mortar (1:3), stretcher bond, facework one side | | |
| half brick thick | m | 90.00 |
| half brick thick, curved | m | 104.00 |
| one brick thick | m | 132.00 |
| one thick, curved | m | 146.00 |
| one and a half brick thick | m | 170.00 |
| two brick thick | m | 206.00 |

| | Unit | Rate £ |
|---|---|---|
| Excavate trench, lay concrete foundation, build wall 1m high in common bricks (basic price £200 per 1000) in cement mortar (1:3), stretcher bond, facework both sides | | |
| half brick thick | m | 92.00 |
| half brick thick, curved | m | 106.00 |
| one brick thick | m | 134.00 |
| one thick, curved | m | 148.00 |
| one and a half brick thick | m | 172.00 |
| two brick thick | m | 208.00 |
| Excavate trench, lay concrete foundation, build wall 1m high in common bricks (basic price £250 per 1000) in cement mortar (1:3), stretcher bond | | |
| half brick thick | m | 94.00 |
| half brick thick, curved | m | 108.00 |
| one brick thick | m | 150.00 |
| one thick, curved | m | 154.00 |
| one and a half brick thick | m | 186.00 |
| two brick thick | m | 210.00 |
| Excavate trench, lay concrete foundation, build wall 1m high in common bricks (basic price £250 per 1000) in cement mortar (1:3), stretcher bond, facework one side | | |
| half brick thick | m | 96.00 |
| half brick thick, curved | m | 110.00 |
| one brick thick | m | 152.00 |
| one thick, curved | m | 156.00 |

| | Unit | Rate £ |
|---|---|---|
| **Brick walling (cont'd)** | | |
| one and a half brick thick | m | 188.00 |
| two brick thick | m | 212.00 |
| Excavate trench, lay concrete foundation, build wall 1m high in common bricks (basic price £350 per 1000) in cement mortar (1:3), stretcher bond, facework one side | | |
| half brick thick | m | 104.00 |
| half brick thick, curved | m | 118.00 |
| one brick thick | m | 152.00 |
| one thick, curved | m | 166.00 |
| one and a half brick thick | m | 190.00 |
| two brick thick | m | 215.00 |
| Excavate trench, lay concrete foundation, build wall 1m high in facing bricks (basic price £350 per 1000) in cement mortar (1:3), English garden wall bond, facework both sides | | |
| half brick thick | m | 108.00 |
| half brick thick, curved | m | 122.00 |
| one brick thick | m | 156.00 |
| one thick, curved | m | 170.00 |
| one and a half brick thick | m | 194.00 |
| two brick thick | m | 220.00 |
| Excavate trench, lay concrete foundation, build wall 1m high in facing bricks (basic price £500 per 1000) in cement mortar (1:3), English garden wall bond, facework one side | | |

| | Unit | Rate £ |
|---|---|---|
| half brick thick | m | 118.00 |
| half brick thick, curved | m | 132.00 |
| one brick thick | m | 170.00 |
| one thick, curved | m | 184.00 |
| one and a half brick thick | m | 240.00 |
| two brick thick | m | 264.00 |
| Excavate trench, lay concrete foundation, build wall 1m high in facing bricks (basic price £500 per 1000) in cement mortar (1:3), English garden wall bond, facework both sides | | |
| half brick thick | m | 120.00 |
| half brick thick, curved | m | 134.00 |
| one brick thick | m | 172.00 |
| one thick, curved | m | 186.00 |
| one and a half brick thick | m | 242.00 |
| two brick thick | m | 260.00 |
| Excavate trench, lay concrete foundation, build wall 1m high in facing bricks (basic price £250 per 1000) in cement mortar (1:3), Flemish bond, facework one side | | |
| half brick thick | m | 100.00 |
| half brick thick, curved | m | 114.00 |
| one brick thick | m | 166.00 |
| one thick, curved | m | 180.00 |

| | Unit | Rate £ |
|---|---|---|
| **Brick walling (cont'd)** | | |
| Excavate trench, lay concrete foundation, build wall 1m high in facing bricks (basic price £250 per 1000) in cement mortar (1:3), Flemish bond, facework both sides | | |
| half brick thick | m | 102.00 |
| half brick thick, curved | m | 116.00 |
| one brick thick | m | 168.00 |
| one thick, curved | m | 182.00 |
| Excavate trench, lay concrete foundation, build wall 1m high in facing bricks (basic price £350 per 1000) in cement mortar (1:3), Flemish bond, facework one side | | |
| half brick thick | m | 108.00 |
| half brick thick, curved | m | 122.00 |
| one brick thick | m | 178.00 |
| one thick, curved | m | 192.00 |
| Excavate trench, lay concrete foundation, build wall 1m high in facing bricks (basic price £350 per 1000) in cement mortar (1:3), Flemish bond, facework both sides | | |
| half brick thick | m | 110.00 |
| half brick thick, curved | m | 124.00 |

| | Unit | Rate £ |
|---|---|---|
| one brick thick | m | 180.00 |
| one thick, curved | m | 194.00 |

Excavate trench, lay concrete foundation, build wall 1m high in facing bricks (basic price £500 per 1000) in cement mortar (1:3), Flemish bond, facework one side

| | | |
|---|---|---|
| half brick thick | m | 130.00 |
| half brick thick, curved | m | 144.00 |
| one brick thick | m | 182.00 |
| one thick, curved | m | 198.00 |

Excavate trench, lay concrete foundation, build wall 1m high in facing bricks (basic price £500 per 1000) in cement mortar (1:3), Flemish bond, facework both sides

| | | |
|---|---|---|
| half brick thick | m | 132.00 |
| half brick thick, curved | m | 146.00 |
| one brick thick | m | 184.00 |
| one thick, curved | m | 198.00 |

| | Unit | Rate £ |
|---|---|---|
| **MASONRY** | | |
| Excavate trench, lay concrete foundation, build random rubble wall laid dry, 1m high, thickness | | |
| 300mm | m | 150.00 |
| 450mm | m | 205.00 |
| 500mm | m | 230.00 |
| Excavate trench, lay concrete foundation, build random rubble wall battered one side, laid dry, 1m high, thickness | | |
| 300mm | m | 160.00 |
| 450mm | m | 220.00 |
| 500mm | m | 245.00 |
| Excavate trench, lay concrete foundation, build random rubble wall battered both sides, laid dry, 1m high, thickness | | |
| 300mm | m | 165.00 |
| 450mm | m | 225.00 |
| 500mm | m | 250.00 |
| Excavate trench, lay concrete foundation, build random rubble wall in gauged mortar (1:1:6), 1m high, thickness | | |
| 300mm | m | 175.00 |
| 450mm | m | 235.00 |
| 500mm | m | 260.00 |

| | Unit | Rate £ |
|---|---|---|
| Excavate trench, lay concrete foundation, build random rubble wall in gauged mortar (1:1:6), 1m high, battered one side, thickness | | |
| 300mm | m | 180.00 |
| 450mm | m | 240.00 |
| 500mm | m | 265.00 |
| Excavate trench, lay concrete foundation, build random rubble wall in gauged mortar (1:1:6), 1m high, battered both sides, thickness | | |
| 300mm | m | 185.00 |
| 450mm | m | 245.00 |
| 500mm | m | 270.00 |
| Excavate trench, lay concrete foundation, build irregular coursed rubble wall in gauged mortar (1:1:6), 1m high, thickness | | |
| 300mm | m | 155.00 |
| 450mm | m | 210.00 |
| 500mm | m | 235.00 |
| Excavate trench, lay concrete foundation, build coursed rubble wall in gauged mortar (1:1:6), 1m high, thickness | | |
| 300mm | m | 165.00 |
| 450mm | m | 220.00 |
| 500mm | m | 245.00 |

| | Unit | Rate £ |
|---|---|---|
| **KERBS AND EDGINGS** | | |
| Excavate trench, lay concrete foundation for precast concrete straight kerb, size | | |
| 127 × 254mm | m | 20.00 |
| 152 × 305mm | m | 25.00 |
| Excavate trench, lay concrete foundation for precast concrete curved kerb, size | | |
| 127 × 254mm | m | 25.00 |
| 152 × 305mm | m | 30.00 |
| Excavate trench, lay concrete foundation for precast concrete straight channel, size | | |
| 127 × 254mm | m | 24.00 |
| Excavate trench, lay concrete foundation for precast concrete curved channel, size | | |
| 127 × 254mm | m | 26.00 |
| Excavate trench, lay concrete foundation for precast concrete straight edging, size | | |
| 51 × 152mm | m | 18.00 |
| 52 × 203mm | m | 18.00 |

| | Unit | Rate £ |
|---|---|---|
| **BEDS AND PAVINGS** | | |
| Excavate by hand, lay sub-base of | | |
| granular material 100mm thick, reinforced concrete bed 100mm thick | m2 | 30.00 |
| granular material 150mm thick, reinforced concrete bed 100mm thick | m2 | 35.00 |
| sand 100mm thick, reinforced concrete bed 100mm thick | m2 | 30.00 |
| sand 150mm thick, reinforced concrete bed 150mm thick | m2 | 38.00 |
| hardcore 100mm thick, reinforced concrete bed 100mm thick | m2 | 42.00 |
| hardcore 150mm thick, reinforced concrete bed 150mm thick | m2 | 48.00 |
| Excavate by hand, lay sub-base of sand 100mm thick, lay brick paving 215 × 103mm (£250 per 1000) | | |
| straight joints both ways | | |
| bricks laid flat | m2 | 60.00 |
| bricks laid on edge | m2 | 75.00 |
| herringbone pattern | | |
| bricks laid flat | m2 | 65.00 |
| bricks laid on edge | m2 | 80.00 |
| Excavate by hand, lay sub-base of sand 100mm thick, lay brick paving 215 × 103mm (£350 per 1000) | | |

| | Unit | Rate £ |
|---|---|---|
| straight joints both ways | | |
| bricks laid flat | m2 | 74.00 |
| bricks laid on edge | m2 | 88.00 |
| herringbone pattern | | |
| bricks laid flat | m2 | 80.00 |
| bricks laid on edge | m2 | 94.00 |
| Excavate by hand, lay sub-base of sand 100mm thick, lay brick paving 215 × 103mm (£500 per 1000) | | |
| straight joints both ways | | |
| bricks laid flat | m2 | 82.00 |
| bricks laid on edge | m2 | 94.00 |
| herringbone pattern | | |
| bricks laid flat | m2 | 86.00 |
| bricks laid on edge | m2 | 98.00 |
| Excavate by hand, lay sub-base of sand 100mm thick, lay precast precast concrete flags, spot bedded cement mortar | | |
| natural colour | | |
| 450 × 450 × 50mm | m2 | 46.00 |
| 450 × 450 × 50mm | m2 | 42.00 |
| 600 × 600 × 50mm | m2 | 38.00 |
| 600 × 900 × 50mm | m2 | 36.00 |
| coloured | | |
| 450 × 450 × 50mm | m2 | 50.00 |
| 450 × 450 × 50mm | m2 | 46.00 |
| 600 × 600 × 50mm | m2 | 42.00 |
| 600 × 900 × 50mm | m2 | 40.00 |

| | Unit | Rate £ |
|---|---|---|
| Excavate by hand, lay sub-base of sand 100mm thick, lay reconstituted York stone flags, spot bedded cement mortar | | |
| 300 × 300 × 50mm | m2 | 60.00 |
| 300 × 450 × 50mm | m2 | 55.00 |
| 300 × 600 × 50mm | m2 | 50.00 |
| 450 × 450 × 50mm | m2 | 48.00 |
| 600 × 600 × 50mm | m2 | 46.00 |
| Excavate by hand, lay sub-base of sand 100mm thick, lay York stone flags, spot bedded in cement mortar, laid straight both ways | | |
| 50mm thick | | |
| cut to size | m2 | 130.00 |
| random sizes | m2 | 115.00 |
| 63mm thick | | |
| cut to size | m2 | 140.00 |
| random sizes | m2 | 125.00 |
| 75mm thick | | |
| cut to size | m2 | 155.00 |
| random sizes | m2 | 135.00 |
| Excavate by hand, lay sub-base of sand 100mm thick, lay natural granite setts, size 100 × 100 × 150mm, bedded in cement mortar | m2 | 95.00 |

| | Unit | Rate £ |
|---|---|---|
| Excavate by hand, lay sub-base of sand 100mm thick, lay concrete setts, bedded in cement mortar, size | | |
| 150 × 150 × 80mm | m2 | 95.00 |
| 200 × 150 × 80mm | m2 | 90.00 |
| Excavate by hand, lay sub-base of sand 100mm thick, lay concrete paviors, bedded in cement mortar, size | | |
| 200 × 100 × 60mm | m2 | 40.00 |
| 200 × 100 × 80mm | m2 | 42.00 |

**DRAINAGE**

| | | |
|---|---|---|
| Excavate trench by hand, lay 100mm diameter 'Supersleve' pipe, granular bed and surround, trench depth | | |
| 0.50m | m | 47.00 |
| 0.75m | m | 57.00 |
| 1.00m | m | 67.00 |
| 1.25m | m | 85.00 |
| 1.50m | m | 112.00 |
| 1.75m | m | 129.00 |
| 2.00m | m | 150.00 |
| 2.25m | m | 167.00 |
| 2.50m | m | 189.00 |
| 2.75m | m | 207.00 |
| 3.00m | m | 234.00 |

| | Unit | Rate £ |
|---|---|---|
| Excavate trench by hand, lay 150mm diameter 'Hepsleve' pipe, granular bed and surround, trench depth | | |
| 0.50m | m | 57.00 |
| 0.75m | m | 67.00 |
| 1.00m | m | 77.00 |
| 1.25m | m | 95.00 |
| 1.50m | m | 122.00 |
| 1.75m | m | 140.00 |
| 2.00m | m | 150.00 |
| 2.25m | m | 177.00 |
| 2.50m | m | 194.00 |
| 2.75m | m | 217.00 |
| 3.00m | m | 244.00 |
| Manhole complete including hand excavation, concrete base and benching, engineering brickwork, channels and cast iron cover, depth | | |
| 1.00m | nr | 525.00 |
| 1.50m | nr | 660.00 |
| 2.00m | nr | 810.00 |

**Land drainage**

| | Unit | Rate £ |
|---|---|---|
| Excavate trench 225mm wide for 75mm clay field drain pipe and backfill with gravel | | |
| trench depth, 500mm | m | 15.00 |
| trench depth, 750mm | m | 18.00 |
| trench depth, 1000mm | m | 22.00 |

| | Unit | Rate £ |
|---|---|---|
| Excavate trench 300mm wide for 75mm clay field drain pipe and backfill with gravel | | |
| trench depth, 500mm | m | 17.00 |
| trench depth, 750mm | m | 21.00 |
| trench depth, 1000mm | m | 25.00 |
| Excavate trench 225mm wide for 100mm clay field drain pipe and backfill with gravel | | |
| trench depth, 500mm | m | 18.00 |
| trench depth, 750mm | m | 21.00 |
| trench depth, 1000mm | m | 25.00 |
| Excavate trench 300mm wide for 100mm clay field drain pipe and backfill with gravel | | |
| trench depth, 500mm | m | 18.00 |
| trench depth, 750mm | m | 21.00 |
| trench depth, 1000mm | m | 25.00 |
| Excavate trench 225mm wide for 100mm clay field drain pipe and backfill with gravel | | |
| trench depth, 500mm | m | 21.00 |
| trench depth, 750mm | m | 24.00 |
| trench depth, 1000mm | m | 28.00 |

| | Unit | Rate £ |
|---|---|---|
| Excavate trench 300mm wide for 100mm clay field drain pipe and backfill with gravel | | |
| trench depth, 500mm | m | 22.00 |
| trench depth, 750mm | m | 25.00 |
| trench depth, 1000mm | m | 29.00 |
| Excavate trench 225mm wide for 60mm perforated field drain pipe and backfill with gravel | | |
| trench depth, 500mm | m | 11.00 |
| trench depth, 750mm | m | 14.00 |
| trench depth, 1000mm | m | 18.00 |
| Excavate trench 300mm wide for 80mm perforated field drain pipe and backfill with gravel | | |
| trench depth, 500mm | m | 13.00 |
| trench depth, 750mm | m | 17.00 |
| trench depth, 1000mm | m | 21.00 |
| Excavate trench 225mm wide for 100mm clay field drain pipe and backfill with gravel | | |
| trench depth, 500mm | m | 14.00 |
| trench depth, 750mm | m | 17.00 |
| trench depth, 1000mm | m | 21.00 |
| Excavate trench 300mm wide for 100mm clay field drain pipe and backfill with gravel | | |

| | Unit | Rate £ |
|---|---|---|
| **Land drainage (cont'd)** | | |
| trench depth, 500mm | m | 16.00 |
| trench depth, 750mm | m | 20.00 |
| trench depth, 1000mm | m | 24.00 |
| Excavate trench 225mm wide for 100mm clay field drain pipe and backfill with gravel | | |
| trench depth, 500mm | m | 19.00 |
| trench depth, 750mm | m | 22.00 |
| trench depth, 1000mm | m | 26.00 |
| Excavate trench 300mm wide for 100mm clay field drain pipe and backfill with gravel | | |
| trench depth, 500mm | m | 21.00 |
| trench depth, 750mm | m | 25.00 |
| trench depth, 1000mm | m | 29.00 |

# Part Five

## TOOL AND EQUIPMENT HIRE

## TOOL AND EQUIPMENT HIRE

The following rates are based on average hire charges made by hire firms in the UK. Check your local dealer for more information. The rates exclude VAT.

### Alloy towers

| | 24 hours £ | Extra 24 hours £ | Week £ |
|---|---|---|---|
| Single width, height | | | |
| 2.2m | 41.00 | 22.00 | 82.00 |
| 3.2m | 51.00 | 26.00 | 102.00 |
| 4.2m | 60.00 | 30.00 | 120.00 |
| 5.2m | 70.00 | 35.00 | 140.00 |
| 6.2m | 78.00 | 39.00 | 156.00 |
| 7.2m | 87.00 | 43.00 | 174.00 |
| 8.2m | 96.00 | 48.00 | 192.00 |
| 9.2m | 104.00 | 52.00 | 208.00 |
| 10.2m | 112.00 | 56.00 | 224.00 |
| Full width, height | | | |
| 2.2m | 76.00 | 38.00 | 152.00 |
| 3.2m | 88.00 | 44.00 | 176.00 |
| 4.2m | 100.00 | 50.00 | 200.00 |
| 5.2m | 110.00 | 55.00 | 220.00 |
| 6.2m | 124.00 | 62.00 | 248.00 |
| 7.2m | 148.00 | 74.00 | 296.00 |
| 8.2m | 160.00 | 80.00 | 320.00 |

| | 24 hours | Extra 24 hours | Week |
|---|---|---|---|
| | £ | £ | £ |
| **Ladders** | | | |
| Roof ladders | | | |
| single, 5.9m | 24.00 | 12.00 | 48.00 |
| double, 4.6m | 22.00 | 11.00 | 44.00 |
| double, 7.6m | 26.00 | 13.00 | 46.00 |
| Push-up ladders | | | |
| double, 3.5m | 16.00 | 8.00 | 32.00 |
| double, 5.0m | 22.00 | 11.00 | 44.00 |
| treble, 2.5m | 16.00 | 8.00 | 32.00 |
| treble, 3.5m | 22.00 | 11.00 | 44.00 |
| Rope-operated ladders | | | |
| double, 6.0m | 38.00 | 19.00 | 76.00 |
| treble, 5.2m | 44.00 | 22.00 | 88.00 |
| treble, 6.0m | 50.00 | 25.00 | 100.00 |
| **Compaction** | | | |
| Vibrating rollers | | | |
| small | 58.00 | 29.00 | 115.00 |
| Rammer | 45.00 | 25.00 | 90.00 |
| Vibrating plate | | | |
| petrol | 26.00 | 13.00 | 52.00 |

| | 24 hours<br>£ | Extra 24 hours<br>£ | Week<br>£ |
|---|---|---|---|
| **Concrete work** | | | |
| Tip up | | | |
| electric | 12.00 | 6.00 | 25.00 |
| petrol | 13.00 | 6.00 | 26.00 |
| Beam screed | 60.00 | 30.00 | 120.00 |
| Power trowel | 41.00 | 20.00 | 82.00 |
| Poker vibrator | | | |
| electric | 34.00 | 17.00 | 68.00 |
| **Pumping** | | | |
| Diaphragm pumps | | | |
| 50mm | 45.00 | 23.00 | 90.00 |
| 75mm | 55.00 | 28.00 | 110.00 |
| Centrifugal pumps | | | |
| 50mm | 28.00 | 14.00 | 56.00 |
| 75mm | 33.00 | 17.00 | 66.00 |
| **Drainage** | | | |
| Drain snake | | | |
| man held | 22.00 | 11.00 | 44.00 |
| Power jet | 170.00 | 85.00 | 340.00 |
| Drain inspection camera | 86.00 | 43.00 | 172.00 |

| | 24 hours | Extra 24 hours | Week |
|---|---|---|---|
| | £ | £ | £ |
| **Landscaping** | | | |
| Lawn aerator | | | |
| powered | 61.00 | 30.00 | 122.00 |
| man held | 10.00 | 5.00 | 20.00 |
| Roller | 10.00 | 5.00 | 20.00 |
| Mower | | | |
| rotary with box (500mm) | 14.00 | 7.00 | 28.00 |
| Scarifier | 37.00 | 18.00 | 74.00 |
| Flame gun | 16.00 | 8.00 | 32.00 |
| Turf cutter | 64.00 | 32.00 | 128.00 |
| Power digger | 41.00 | 20.00 | 52.00 |
| Tiller | 22.00 | 11.00 | 44.00 |
| Stump grinder | 99.00 | 50.00 | 198.00 |
| Shredder | | | |
| 35mm | 19.00 | 10.00 | 38.00 |

| | 24 hours | Extra 24 hours | Week |
|---|---|---|---|
| | £ | £ | £ |
| Hedge trimmer | | | |
| electric | 12.00 | 6.00 | 24.00 |
| petrol | 31.00 | 15.00 | 62.00 |
| Long handled trimmer | | | |
| 1.3m | 39.00 | 20.00 | 78.00 |
| Logging saw | 34.00 | 17.00 | 68.00 |
| Pruning saw | 15.00 | 8.00 | 30.00 |
| Long handle pruner | | | |
| petrol | 42.00 | 21.00 | 84.00 |
| man held | 8.00 | 4.00 | 16.00 |
| Chain saw | | | |
| petrol | 60.00 | 20.00 | 98.00 |
| Leaf sucker | | | |
| petrol | 25.00 | 13.00 | 50.00 |
| Fence post borer | | | |
| petrol | 42.00 | 21.00 | 84.00 |
| man held | 8.00 | 4.00 | 16.00 |

**Sundries**

| | | | |
|---|---|---|---|
| Tarpaulins | | | |
| 4 × 5m | | | 22.00 |
| 8 × 5m | | | 28.00 |

# Part Six

## BUSINESS MATTERS

Starting a business

Running a business

Taxation

## STARTING A BUSINESS

Most small businesses come into being for one of two reasons – ambition or desperation! A person with genuine ambition for commercial success will never be completely satisfied until he has become self-employed and started his own business. But many successful businesses have been started because the proprietor was forced into this course of action because of redundancy.

Before giving up his job, the would-be businessman should consider carefully whether he has the required skills and the temperament to survive in the highly competitive self-employed market. Before commencing in business it is essential to assess the commercial viability of the intended business because it is pointless to finance a business that is not going to be commercially viable.

In the early stages it is important to make decisions such as: What exactly is the product being sold? What is the market view of that product? What steps are required before the developed product is first sold and where are those sales coming from?

As much information as possible should be obtained on how to run a business before taking the plunge. Sales targets should be set and it should be clearly established how those important first sales are obtained. Above all, do not underestimate the amount of time required to establish and finance a new business venture.

Whatever the size of the business it is important that you put in writing exactly what you are trying to do. This means preparing a business plan that will not only assist in establishing your business aims but is essential if you need to raise finance. The contents of a typical business plan are set out later. It is important to realise that you are not on your own and there are many contacts and advising agencies that can be of assistance.

### Potential customers and trade contacts

Many persons intending to start a business in the construction industry will have already had experience as employees. Use all contacts to check the market, establish the sort of work that is available and the current charge out rates.

In the domestic market, check on the competition for prices and services provided. Study advertisements for your kind of work and try to get firm promises of work before the start-up date.

## Testing the market

Talk to as many traders as possible operating in the same field. Identify if the market is in the industrial, commercial, local government or in the domestic field. Talk to prospective customers and clients and consider how you can improve on what is being offered in terms of price, quality, speed, convenience, reliability and back-up service.

## Business links

There is no shortage of information about the many aspects of starting and running your own business. Finance, marketing, legal requirements, developing your business idea and taxation matters are all the subject of a mountain of books, pamphlets, guides and courses so it should not be necessary to pay out a lot of money for this information. Indeed, the likelihood is that the aspiring businessman will be overwhelmed with information and will need professional guidance to reduce the risk of wasting time on studying unnecessary subjects.

Business Links are now well established and provide a good place to start for both information and advice. These organisations provide a 'one-stop-shop' for advice and assistance to owner-managed businesses. They will often replace the need to contact Training and Enterprise Councils (TECs) and many of the other official organisations listed below.
Point of contact: telephone directory for address.

## Training and Enterprise Councils (TECs)

TECs are comprised of a board of directors drawn from the top men in local industry, commerce, education, trade unions etc., who, together with their staff and experienced business counsellors, assist both new and established concerns in all aspects of running a business. This takes the form of across-the-table advice and also hands-on assistance in management, marketing and finance if required. There are also training courses and seminars available in most areas together with the possibility of grants in some areas.
Point of contact: local Jobcentre or Citizens' Advice Bureau.

## Banks

Approach banks for information about the business accounts and financial services that are available. Your local Business Link can advise on how best to find a suitable bank manager and inform you as to what the bank will require.

Shop around several banks and branches if you are not satisfied at first because managers vary widely in their views on what is a viable business proposition. Remember, most banks have useful free information packs to help business start-up.

Point of contact: local bank manager.

## HM Inspector of Taxes

Make a preliminary visit to the local tax office enquiry counter for their publications on income tax and national insurance contributions.

| | |
|---|---|
| SA/Bk 3 | Self assessment. A guide to keeping records for the self employed |
| IR 15(CIS) | Construction Industry Tax Deduction Scheme |
| CWL | Starting your own business, |
| IR 40(CIS) | Conditions for Getting a Sub-Contractor's Tax Certificate |
| NE1 | PAYE for Employers (if you employ someone) |
| NE3 | PAYE for new and small Employers |
| IR 56/N139 | Employed or Self-Employed. A guide for tax and National Insurance |
| CA02 | National Insurance contributions for self employed people with small earnings. |

Remember, the onus is on the taxpayer, within three months, to notify the Inland Revenue that he is in business and failure to do so may result in the imposition of £100 penalty. Either send a letter or use the form provided at the back of the *'Starting your own business booklet'* to the Inland Revenue National Insurance Contributions Office and they will inform your local tax office of the change in your employment status.

Point of contact: telephone directory for address.

### Inland Revenue National Insurance Contributions Office

Self Employment Services
Customer Accounts Section
Longbenton
Newcastle NE 98 1ZZ

Telephone the Call Centre on 0845 9154655 and ask for the following publications:

| | |
|---|---|
| CWL2 | Class 2 and Class 4 Contributions for the Self Employed |
| CA02 | People with Small Earnings from Self-Employment |
| CA04 | Direct Debit – The Easy Way to Pay. Class 2 and Class 3 |
| CA07 | Unpaid and Late Paid Contributions and for Employers |
| CWG1 | Employer's Quick Guide to PAYE and NIC Contributions |
| CA30 | Employer's Manual to Statutory Sick Pay |

### VAT

The VAT office also offer a number of useful publications, including;

| | |
|---|---|
| 700 | The VAT Guide |
| 700/1 | Should I be Registered for VAT? |
| 731 | Cash Accounting |
| 732 | Annual Accounting |
| 742 | Land and Property |

Information about the Cash Accounting Scheme and the introduction of annual VAT returns are dealt with later.
Point of contact: telephone directory for address.

### Local authorities

Authorities vary in provisions made for small businesses but all have been asked to simplify and cut delays in planning applications. In Assisted Areas, rent-free periods and reductions in rates may be available on certain industrial and commercial properties. As a preliminary to either purchasing or renting business premises, the following booklets will be helpful:

*Step by Step Guide to Planning Permission for Small Businesses,* and
*Business Leases and Security of Tenure*

Both are issued by the Department of Employment and are available at council offices, Citizens' Advice Bureau and TEC offices. Some authorities run training schemes in conjunction with local industry and educational establishments.

Point of contact: usually the Planning Department - ask for the Industrial Development or Economic Development Officer.

**Department of Trade and Industry**

The services formally provided by the Department are now increasingly being provided by Business Link. The Department can still, however, provide useful information on available grants for start-ups.
Point of contact: telephone 0207-215 5000 and ask for the address and telephone number of the nearest DTI office and copies of their explanatory booklets.

**Department of Transport and the Regions**

Regulations are now in force relating to all forms of waste other than normal household rubbish. Any business that produces, stores, treats, processes, transports, recycles or disposes of such waste has a 'duty of care' to ensure it is properly discarded and dealt with.

Practical guidance on how to comply with the law (it is a criminal offence punishable by a fine not to) is contained in a booklet *Waste Management: The Duty of Care: A Code of Practice* obtainable from HMSO Publication Centre, PO Box 276, London SW8 5DT. Telephone 0207-873 9090.

**Accountant**

The services of an accountant are to be strongly recommended from the beginning because the legal and taxation requirements start immediately and must be properly complied with if trouble is to be avoided later. A qualified accountant must be used if a limited company is being formed but an accountant will give advice on a whole range of business issues including book-keeping, tax planning and compliance to finance raising and will help in preparing annual accounts.

It is worth spending some time finding an accountant who has other clients in the same line of business and is able to give sound advice particularly on taxation and business finance and is not so overworked that damaging delays in producing accounts are likely to arise. Ask other traders whether they can recommend their own accountant. Visit more than one firm of accountants, ask about the fees they charge and how much the production of annual accounts and agreement with the Inland Revenue are likely to cost. A good accountant is worth every penny of his fees and will save you both money and worry.

## Solicitor

Many businesses operate without the services of a solicitor but there are a number of occasions when legal advice should be sought. In particular, no one should sign a lease of premises without taking legal advice because a business can encounter financial difficulty through unnoticed liabilities in its lease. Either an accountant or solicitor will help with drawing up a partnership agreement that all partnerships should have. A solicitor will also help to explain complex contractual terms and prepare draft contracts if the type of business being entered into requires them.

## Insurance broker

Policies are available to cover many aspects of business including:

- employer's liability – compulsory if the business has employees
- public liability – essential in the construction industry
- motor vehicles
- theft of stock, plant and money
- fire and storm damage
- personal accident and loss of profits
- key man cover.

Brokers are independent advisers who will obtain competitive quotations on your behalf. See more than one broker before making a decision – their advice is normally given free and without obligation.
Point of contact: telephone directory or write for a list of local members to:

The British Insurance Brokers' Association
Consumer Relations Department
BIBA House
14 Bevis Marks
London
EC3A 7NT (telephone: 0207-623 9043)

or contact
The Association of British Insurers
51 Gresham Street
London
EC2V 7HQ (telephone: 0207-600 3333)

who will supply free a package of very useful advice files specially designed for the small business.

**The Health and Safety Executive**

The Executive operates the legislation covering everyone engaged in work activities and has issued a very useful set of '*Construction Health Hazard Information Sheets*' covering such topics as handling cement, lead and solvents, safety in the use of ladders, scaffolding, hoists, cranes, flammable liquids, asbestos, roofs and compressed gases etc. A pack of these may be obtained free from your local HSE office or The Health & Safety Executive Central Office, Sheffield (telephone: 01142-892345) or HSE Publications (telephone: 01787-881165).

**Business plan**

As stated before, once the relevant information has been obtained it should be consolidated into a formal business plan. The complexity of the plan will depend in the main on the size and nature of the business concerned. Consideration should be given to the following points.

## Objectives

It is important to establish what you are trying to achieve both for you and the business. A provider of finance may be particularly influenced by your ability to achieve short- and medium-term goals and may have confidence in continuing to provide finance for the business. From an individual point of view, it is important to establish goals because there is little point in having a business that only serves to achieve the expectations of others whilst not rewarding the would-be businessman.

### History

If you already own an existing business then commentary on its existing background structure and history to date can be of assistance. There is no substitute for experience and any existing contacts you have in the construction industry will be of assistance to you. The following points should also be considered for inclusion:

- a brief history of the business identifying useful contacts made
- the development of the business, highlighting significant successes and their relevance to the future
- principal reasons for taking the decision to pursue this new venture
- details of present financing of the business.

### Products or services

It is important to establish precisely what it is you are going to sell. Does the product or service have any unique qualities which gives it your advantages over competitors? For example, do you have an ability to react more quickly than your competitors and are you perceived to deliver a higher quality product or service? A typical business plan would include:

- description of the main products and services
- statement of disadvantages and advising how they will be overcome
- details of new products and estimated dates of introduction

- profitability of each product
- details of research and development projects
- after-sales support.

Markets and marketing strategy

This section of the business plan should show that thought has been given to the potential of the product. In this regard it can often be useful to identify major competitors and make an overall assessment of their strengths and weaknesses, including the following:

- an overall assessment of the market, setting out its size and growth potential
- a statement showing your position within the market
- an identification of main customers and how they compare
- details of typical orders and buying habits
- pricing strategy
- anticipated effect on demand of pricing
- expectation of price movement
- details of promotions and advertising campaigns.

It is important to identify your customers and why they might buy from you. Those entering the domestic side of the business will need to think about the best way to reach potential customers. Are local word-of-mouth recommendations enough to provide reasonable work continuity. If not, what is the most effective method of advertising to reach your customer base?

Remember, advertising is costly. It is a waste of funds to place an advertisement in a paper circulating in areas A, B, C & D if the business only covers area A.

Research and development

If you are developing a product or a particular service, then an assessment should be made on what stage it is at and what further finance is required to complete it. It may also be useful to make an assessment on the vulnerability of the product or service to innovations being initiated by others.

## Basis of operation

Detail what facilities you will require in order to carry on your trade in the form of property, working and storage areas, office space, etc. An assessment should also he made on the assistance you will require from others. Your business plan might include:

- a layman's guide to the process or work
- details of facilities, buildings and plant
- key factors affecting production, such as yields and wastage
- raw material demand and usage.

## Management

This section is one of the most important because it demonstrates the capability of the would-be businessman. The skills you need will cover production, marketing, finance and administration. In the early stages you may be able to do this yourself but as the business grows it may be required to develop a team to handle these matters. The following points should be considered for inclusion in the plan:

- set out age, experience and achievements
- state additional management requirements in the future and how they are to be met
- identify current weaknesses and how they will be overcome
- state remuneration packages and profit expectations
- give detailed CVs in appendices.

Advertising and retraining may be required in order to identify and provide suitable personnel where expertise and experience are lacking.

## Financial information

It is important to detail, if any, the present financial position of your business and the budgeted profit and loss accounts, cash flows and balance sheets. These integrated forecasts should be prepared for the next twelve months at monthly intervals and annually for the following two years.

If the forecasts are to be reasonably accurate then the businessman must make some early decisions about:

- the premises where the business will be based, the initial repairs and alterations that might he required and an assessment of the total cost
- which plant, equipment and transport are needed, whether they are to be leased or purchased and what the cost will be?
- how much stock of materials, if any, should be carried? the bare minimum only should be acquired, so reliable suppliers should be found
- what will be the weekly bills for overheads, wages and the proprietor's living costs?
- what type of work is going to be undertaken, and how much profit can realistically be obtained?
- how often are invoices to be presented?

Your business plan should include the following information:

- explanation of how sales forecasts are prepared
- levels of production
- details of major variable overheads and estimates
- assumptions in cash flow forecasting, inflation and taxation.

Finance required and its application

The financial details given above should produce an accurate assessment of the funds required to finance the business. It is important to distinguish between those items that require permanent finance and those that will eventually be converted to cash because it is not usually advisable to finance long-term assets with personal equity.

Working capital such as stock and debtors can usually be obtained by an overdraft arrangement but your accountant or bank will advise you on this.

Executive summary

Although it is prepared last, this summary will be the first part of your business plan. Remember that business plans are prepared for busy people and their decision on finance may be based solely on this section. It should cover two or three pages and deal with the most important aspects and opportunities in your plan. Here are some of the main headings:

- key strategies
- finance required and how it is to be used
- management experience
- anticipated returns and profits
- markets.

The appendices should include:

- CVs of key personnel
- organisation charts
- market studies
- product advertising literature
- professional references
- financial forecasts
- glossary of terms.

If you feel that any additional information should be provided in support of your proposal, then this is usually best included in the appendices.

### Follow up

Please remember that once your plan is prepared, it is important to examine it again regularly and update the forecasts and financial information. This is a working document and can be an important tool in running the business.

## Sources of finance

### Personal funds

Finance, like charity, often begins at home and a would-be businessman should make a realistic assessment of his net worth, including the value of his house after deducting the mortgage(s) outstanding on it, savings, any car or van owned and any sums which the family are prepared to contribute but deducting any private borrowings which will come due for payment. The whole of these funds may not be available (for instance, money which has been loaned to a friend or relative who is known to be unable to repay at the present time).

It may not be desirable that all capital should be put at risk on a business venture so the following should be established:

- how much cash you propose to invest in the business
- whether the family home will be made available for any business borrowing
- state total finance required
- how finance is anticipated being raised
- interest and security to be provided
- expected return on investment.

Whilst it may be wise not to pledge too much of the family assets, it has to be remembered that the bank will be looking closely at the degree to which the proprietor has committed himself to the venture and will not be impressed by an application for a loan where the applicant is prepared to risk only a small fraction of his own resources.

Having decided how much of his own funds to contribute, the businessman can now see the level of shortfall and consider how best to fill it. Consideration should be given to partners where the shortfall is large and particularly when there is a need for heavy investment in fixed assets, such as premises and capital equipment. It may be worthwhile starting a limited company with others also subscribing capital and to allow the banks to take security against the book debts.

Banks

The first outside source of money to which most businessmen turn is the bank and here are a few guidelines on approaching a bank manager:

- present your business plan to him; remember to use conservative estimates which tend to understate rather than overstate the forecast sales and profits
- know the figures in detail and do not leave it to your accountant to explain them for you. The bank manager is interested in the businessman not his advisers and will be impressed if the businessman demonstrates a grasp of the financing of his business

- understand the difference between short- and long-term borrowing
- ask about the Government Loan Guarantee Scheme if there is a shortage of security for loans. The bank may be able to assist, or depending on certain conditions being met, the Government may guarantee a certain percentage of the bank loan.

Remember the bank will want their money back, so bank borrowings are usually required to be secured by charges on business assets. In start-up situations, personal guarantees from the proprietors are normally required. Ensure that if these are given they are regularly reviewed to see if they are still required.

Enterprise Investment Scheme – business angels

If an outside investor is sought in a business he will probably wish to invest within the terms of the Enterprise Investment Scheme which enables him to gain income tax relief at 20% on the amount of his investment. Additionally, any investment can be used to defer capital gains tax. The rules are complex and professional advice should always be sought.

Hire purchase/leasing

It is not always necessary to purchase assets outright that are required for the business and leasing and hire purchase can often form an integral part of a business's medium-term finance strategy.

Venture capital

In addition, there are a number of other financial institutions in the venture capital market that can help well-established businesses, usually limited companies, who wish to expand. They may also assist well-conceived start-ups. They will provide a flexible package of equity and loan capital but only for large amounts, usually sums in excess of £150,000 and often £250,000.

Usually the deal involves the financial institution having a minority interest in the voting share capital and a seat on the board of the company. Arrangements for the eventual purchase of the shares held by the finance company by the private shareholders are also normally incorporated in the scheme.

## The Royal Jubilee and Princes Trust

These trusts through the Youth Business Initiative provide bursaries of not more than £1,000 per individual to selected applicants who are unemployed and age 25 or over. Grants may be used for tools and equipment, transport, fees, insurance, instruction and training but not for working capital, rent and rates, new materials or stock. They operate through a local representative whose name and address may be ascertained by contacting the Prince's Youth Business.
Point of contact: telephone 0207-321 6500.

## The Business Start-up Scheme

This is an allowance of £50 per week, in addition to any income made from your business, paid for twenty weeks. To qualify you must be at least 18 and under 65, work at least 36 hours per week in the business and have been unemployed for at least six months or fall into one of the other categories: disabled, ex-HMS or redundant.

The first step is to get the booklet on the subject from your local Jobcentre or TEC that includes details on how and where to apply. Once in receipt of the enterprise allowance, you will also have the benefit of advice and assistance from an experienced businessman from your TEC. All the initial counselling services and training courses are free.

## RUNNING A BUSINESS

Many businesses are run without adequate information being available to check trend in their vital areas, e.g. marketing, money and managerial efficiency. It is essential to look critically at all aspects of the business in order to maximise profits and reduce inefficiency. Regular meaningful information is required on which management can concentrate. This will vary according to the proprietor's business but will often concentrate on debtors, creditors, cash, sales and orders.

Proprietors often have the feeling that the business should be 'doing better' but are unable to identify what is going wrong. Sometimes there is the worrying phenomenon of a steadily increasing work programme coupled with a persistently reducing bank balance or rising overdraft. Some useful ways of checking the position and of identifying problem areas are given below.

### Marketing

Throughout his business life the entrepreneur should continuously study the methods and approach of his competitors. A shortcoming frequently found in ailing concerns is that the proprietor thinks he knows what his customers want better than they do.

The term 'market research' sounds both difficult and expensive but a very simple form of it can be done quite effectively by the businessman and his sales staff. Existing and prospective customers should be approached and asked what they want in terms of price, quality, design, payment terms, follow-up service, guarantees and services.

The initial approach might be by a leaflet or letter followed by a personal call. As an on-going part of management, all staff with customer contact should be encouraged to enquire about and record customer preferences, complaints, etc. and feed it back to management.

Other sources of information can be trade and business journals, trade exhibitions, suppliers and representatives from which information about trends, new techniques and products can be obtained and studied. Valuable information can also be gained from studying competitors and the following questions should be asked:

- what do they sell and at what prices?
- what inducements do they offer to their customers, e.g. credit facilities, guarantees, free offers and discounts?
- how do they reach their customers - local/national advertising, mail shots, salesmen, local radio and TV?
- what are the strongest aspects of their appeal to customers and have they any weaknesses?

The businessman should apply all the information gathered from customers and competitors to his own services with a view to making sure he is offering the right product at the right price in the most attractive way and in the most receptive market.

In a small business where the proprietor is also his own salesman he must give careful thought on how he can best present his product and himself. For instance, if he is working solely within the construction industry his main problems are likely to centre on getting a C1S6 Certificate and using trade contacts to get sub-contract work.

However, for those who serve the general public, presentation can be a vital element in getting work. The customer is looking for efficiency, reliability and honesty in a trader and quality, price and style in the product. To bring out these facets in discussion with a potential customer is a skilled task. A short course on marketing techniques could pay handsome dividends. The Business Link will give the names and addresses of such courses locally.

## Financial control

Unfortunately, some unsuccessful firms do not seek financial advice until too late when the downward trend cannot be halted. Earlier attention to the problems may have saved some of them so it is important to recognise the tell-tale signs. There are some tests and checks that can be done quite easily.

## Cash flow

Cash flow is the lifeblood of the business and more businesses fail through lack of cash than for any other reason. Cash is generated through the conversion of work into debtors and then into payment and also through the deferral of the payment of supplies for as long a period that can be negotiated. The objective must be to keep stock, work in progress, debts to a minimum and creditors to a maximum.

### Debtor days

This is calculated by dividing your trade debtors by annual sales and multiplying by 365. This shows the number of days' credit being afforded to your customers and should be compared both with your normal trade terms and the previous month's figures. Normal procedures should involve the preparation of a monthly-aged list of debtors showing the name of the customer, the value and to which month it relates.

The oldest and largest debtors can be seen at a glance for immediate consideration of what further recovery action is needed. The list may also show over-reliance on one or two large customers or the need to stop supplying a particularly bad payer until his arrears have been reduced to an acceptable level. Consideration should be given to making up bills to a date before the end of the month and making sure the accounts are sent out immediately, followed by a statement four weeks later.

Consider giving discounts for prompt payment. If all else fails, and legal action for recovery is being contemplated, call at the County Court and ask for their leaflets.

### Stock turn

The level of stock should be kept to a minimum and the number of days' stock can be calculated by dividing the stock by the annual purchases and multiplying by 365. A worsening trend on a month-by-month basis shows the need for action. It is important to regularly make a full inventory of all stock and dispose of old or surplus items for cash. A stock control procedure to avoid stock losses and to keep stock to a minimum should be implemented.

### Profitability

Whilst cash is vital in the short-term, profitability is vital in the medium-term. The two key percentage figures are the gross profit percentage and the net profit percentage. Gross profit is calculated by deducting the cost of materials and direct labour from the sales figures whilst net profit is arrived at after deducting all overheads. Possible reasons for changes in the gross profit percentage are:

- not taking full account of increases in materials and wages in the pricing of jobs
- too generous discount terms being offered
- poor management, over-manning, waste and pilferage of materials
- too much down-time on equipment which is in need of replacement.

If net profit is deteriorating after the deduction of an appropriate reward for your own efforts, including an amount for your own personal tax liability, you should review each item of overhead expenditure in detail asking the following questions:

- can savings be made in non-productive staff?
- is sub-contracting possible and would it be cheaper?
- have all possible energy-saving methods been fully explored?
- do the company's vehicles spend too much time in the yard and can they be shared or their number reduced?
- is the expenditure on advertising producing sales - review in association with 'marketing' above?

**Over-trading**

Many inexperienced businessmen imagine that profitability equals money in the bank and in some cases, particularly where the receipts are wholly in cash, this may be the case. But often, increased business means higher stock inventories, extra wages and overheads, increased capital expenditure on premises and plant, all of which require short-term finance.

Additionally, if the debtors show a marked increase as the turnover rises, the proprietor may find to his surprise that each expansion of trade reduces rather than increases his cash resources and he is continually having to rely on extensions to his existing credit.

The business, which had enough funds for start-up, finds it does not have sufficient cash to run at the higher level of operation and the bank manager may he getting anxious about the increasing overdraft. It is essential for those who run a business that operates on credit terms to be aware that profitability does not necessarily mean increased cash availability. Regular monthly management information on marketing and finance as described in this chapter will enable over-trading to be recognised and remedial action to be taken early.

If the situation is appreciated only when the bank and other creditors are pressing for money, radical solutions may be necessary, such as bringing in new finance, sale and leaseback of premises, a fundamental change in the terms of trade or even selling out to a buyer with more resources. Help from the firm's accountant will be needed in these circumstances.

## Break-even point

The costs of a business may be divided into two types - variable and fixed. *Variable costs* are those which increase or decrease as the volume of work goes up or down and include such items as materials used, direct labour and power machine tools. *Fixed costs* are not related to turnover and are sometimes called fixed overheads. They include rent, rates, insurance, heat and light, office salaries and plant depreciation. These costs are still incurred even though few or no sales are being made.

Many small businessmen run their enterprises from home using family labour as back-up; they mainly sell their own labour and buy materials and hire plant only as required. By these means they reduce their fixed costs to a minimum and start making profits almost immediately. However, larger firms that have business premises, perhaps a small workshop, an office and vehicles, need to know how much they have to sell to cover their costs and become profitable.

In the case of a new business it is necessary to estimate this figure but where annual accounts are available a break-even chart based on them can be readily prepared. Suppose the real or estimated figures (expressed in £000s) are:

| | % | £ |
|---|---|---|
| Sales | 100 | 400 |
| Variable costs | 66 | 265 |
| Gross profit | 34 | 135 |
| Fixed costs | 13 | 50 |
| Net profit | 21 | 85 |

Break-even point = $\frac{\text{50 divided by (1 less variable costs \%)}}{\text{sales}}$

= 50 divided by (1 less 0.6625)
= 50 divided by 0.3375
= £148 (thousand)

In practice, things are never quite as clear cut as the figures show, but nevertheless this is a very useful tool for assessing not only the break-even point but also the approximate amount of loss or profit arising at differing levels of turnover and also for considering pricing policy.

## TAXATION

The first decision usually required to be made from a taxation point of view is which trading entity to adopt. The options available are set out below.

### Sole trader

A sole trader is a person who is in business on his own account. There is no statutory requirement to produce accounts nor is there a necessity to have them audited. A sole trader may, however, be required to register for PAYE and VAT purposes and maintain records so that Income Tax and VAT returns can be made. A sole trader is personally liable for all the liabilities of his business.

### Partnership

A partnership is a collection of individuals in business on their own account and whose constitution is generally governed by the Partnership Act 1890. It is strongly recommended that a partnership agreement is also established to determine the commercial relationship between the individuals concerned.

The requirements in relation to accounting records and returns are similar to those of a sole trader and in general a partner's liability is unlimited.

### Limited company

This is the most common business entity. Companies are incorporated under the Companies Act 1985 which requires that an annual audit is carried out for all companies with a turnover in excess of £5,000,000 or a review if the turnover is less than £5,000,000 and that accounts are filed with the Companies Registrar. Generally an individual shareholder's liability is limited to the amount of the share capital he is required to subscribe.

### Advantages

In view of the problems and costs of incorporating an existing business, it is important to try and select the correct trading medium at the commencement of operations. It is not true to say that every business should start life as a company.

Many businesses are carried on in a safe and efficient manner by sole traders or partnerships. Whilst recognising the possible commercial advantages of a limited company, taxation advantages exist for sole traderships and partnerships, such as income tax deferral and National Insurance saving. No decision should be taken without first seeking professional advice.

The benefit of limited liability should not be ignored although this can largely be negated by banks seeking personal guarantees. In addition, it may be easier for the companies to raise finance because the bank can take security on the debts of the company that could be sold in the future, particularly if third-party finance has been obtained in the form of equity.

## Self-assessment

From the tax year 1996/97 the burden of assessing tax shifted from the Inland Revenue to the individual tax payer. The main features of this system are as follows:

- the onus is on the taxpayer to provide information and to complete returns
- tax will be payable on different dates
- the taxpayer has a choice: he can calculate his tax liability at the same time as making his return and this will need to be done by 31st January following the end of the tax year. Alternatively, he can send in his tax return before 30 September and the Inland Revenue will calculate the tax to be paid on the following 31 January
- the important aspect to the system is that if the return is late, or the tax is paid late, there will be automatic penalties and/or surcharges imposed on the taxpayer.

## Tax correspondence

Businessmen do not like letters from the Inland Revenue but they should resist the temptation to tear them up or put them behind the clock and forget about them. All Tax Calculations and Statements of Account should be checked for accuracy immediately and any queries should be put to your accountant or sent to the Tax District that issued the document.

Keep copies of all correspondence with the Inland Revenue. Letters can be mislaid or fail to be delivered and it is essential to have both proof of what was sent as well as a permanent record of all correspondence.

## Dates tax due

*Income Tax*

Payments on account (based on one half of last year's liability) are due on 31 January and 31 July. If these are insufficient there is a balancing payment due on the following 31 January – the same day as the tax return needs to be filed. For example:

| | |
|---|---|
| for the year 2006/07 | Tax due £5,000 (2005/06 was £4,000)<br>First payment on account of £2,000 is due on 31.01.07<br>Second payment on account of £2,000 is due on 31.07.07<br>Balancing payment of £1,000 is due on 31.01.08 |

Note that on 31.01.07, the first payment on account of £2,500 fell due for the tax year 2007/08.

## Tax in business

*Spouses in business*

If spouses work in the business, perhaps answering the phone, making appointments, writing business letters, making up bills and keeping the books, they should be properly remunerated for it. Being a payment to a family member, the Inspector of Taxes will be understandably cautious in allowing remuneration in full as a business expense. The payment should be:

- actually paid to them, preferably weekly or monthly and in addition to any housekeeping monies
- recorded in the business book

- reasonable in amount in line with their duties and the time spent on them.

If the wages paid to them exceed £96.00 per week, Class 1 employer's and employee's NIC becomes due and if they exceed £5,225 p.a. (assuming they have no other income) PAYE tax will also be payable.

It should also be noted that once small businesses are well established and the spouses' earnings are approaching the above limits, consideration may be given to bringing them in as a partner. This has a number of effects:

- there is a reduced need to relate the spouse's income (which is now a share of the profits) to the work they do
- they will pay Class 2 and Class 4 NIC instead of the more costly Class I contributions and PAYE will no longer apply to their earnings but remember that, as partners, they have unlimited liability.

## Premises

Many small businessmen cannot afford to rent or buy commercial premises and run their enterprises from home using part of it as an office where the books and vouchers, clients' records and trade manuals are kept and where estimates and plans are drawn up. In these circumstances, a portion of the outgoings on the property may be claimed as a business expenses. An accountant's advice should be sought to ensure that the capital gains tax exemption that applies on the sale of the main residence is not lost.

## Fixed Profit Car Scheme

It may be advantageous to calculate your car expenses using a fixed rate per business mile. Ask your accountant about this. A proper record of business mileage must be kept.

## Vehicles

Car expenses for sole traders and partners are usually split on a fractional mileage basis between business journeys, which are allowable, and private ones, which are not, and a record of each should he kept. If the business does work only on one or two sites for only one main contractor, the inspector may argue that the true base of operations is the work site not the residence and seek to disallow the cost of travel between home and work. It is tax-wise and sound business practice to have as many customers as possible and not work for just one client.

### Business entertainment

No tax relief is due for expenditure on business entertainment and neither is the VAT recoverable on gifts to customers, whether they are from this country or overseas. However, the cost of small trade gifts not exceeding £50 per person per annum in value is still admissible provided that the gift advertises the business and does not consist of food, drink or tobacco.

### Income tax (2006/07)

*Personal allowances*

The current personal allowance for a single person is £5,035. The personal allowance for people aged 65 to 74 and over 75 years are £7,280 and £7,420 respectively. The married couple's allowance was withdrawn on 5 April 2000, except for those over 65 on that date.

*Taxation of husband and wife*

A married woman is treated in much the same way as a single person with her own personal allowance and basic rate band. Husband and wife each make a separate return of their own income and the Inland Revenue deals with each one in complete privacy; letters about the husband's affairs will be addressed only to him and about the wife's only to her unless the parties indicate differently.

### Rates of tax

Tax is deducted at source from most banks and building societies accounts at the rate of 20%. The rates of tax for 2006/07 are as follows:

Lower rate: 10% on taxable income up to £2,150
Basic rate: 22% on taxable income between £2,150 and £33,300
Higher rate: 40% on taxable income over £33,300

Dividends carry a 10% non-repayable tax credit. Higher rate taxpayers pay a further tax on dividends of 22.5%.

### Mortgage interest relief

This is no longer available after 5 April 2000.

### Business losses

These are allowed only against the income of the person who incurs the loss. For example, a loss in the husband's business cannot be set against the wife's income from employment.

### Joint income

In the case of joint ownership by a husband and wife of assets that yield income, such as bank and building society accounts, shares and rented property, the Inland Revenue will treat the income as arising equally to both and each will pay tax on one half of the income. If, however, the asset is owned in unequal shares or one spouse only and the taxpayer can prove this, then the shares of income to be taxed can be adjusted accordingly if a joint declaration is made to the tax office setting out the facts.

### Capital Gains Tax

Where an asset is disposed of, the first £8,800 of the gain is exempt from tax. In the case of husbands and wives, each has a £8,800 exemption so if the ownership of the assets is divided between them, it is possible to claim exemption on gains up to £17,600 jointly in the tax year. Any remaining gain is chargeable as though it were the top slice of the individual's income; therefore according to his or her circumstances it might be charged at 10%, 22% or 40%.

**Self-employed NIC rates (from 6 April 2006)**

*Class 2 rate*
Charged at £2.10 per week. If earnings are below £5,035 per annum averaged over the year, ask the DSS about 'small income exception'. Details are in leaflet CA02.

*Class 4 rate*
Business profits up to £5,035 per annum are charged at NIL. Annual profits between £5,035 and £33,540 are charged at 8% of the profit. There is also a charge on profits over £33,540 of 1%. Class 4 contributions are collected by the Inland Revenue along with the income tax due.

*Capital allowances (depreciation) rates*

| | | |
|---|---|---|
| Plant and machinery: | | 25% (40% first-year allowance is available for certain small businesses) |
| Business motor cars | - cost up to £12,000: | 25% |
| | - cost over £12,000: | £3,000 (maximum) |

THE CONSTRUCTION INDUSTRY TAX DEDUCTION SCHEME

**General**

The new Construction Industry Tax Deduction Scheme is known as the 'revised CIS' scheme and replaced the old 'CIS' scheme. Everyone who carries out work in the Construction Industry Scheme must hold a registration card (CIS4) or a tax certificate (CIS6). Certain larger companies use a special certificate (CIS5).

A small business that does work only for the general public and small commercial concerns is outside the scheme and does not need a certificate to trade. If, however, it engages other contractors to do jobs for it, the business would have to trade under the scheme as a contractor and deduct tax from any payment made to a subcontractor who did not produce a valid (CIS6) certificate. If in doubt, consult your accountant or the Inland Revenue direct.

Under the revised scheme, registration cards, tax certificates and vouchers will no longer be used. Subcontractors will normally register with the Revenue. There will be two types of registration, registration for gross payment, applicable to those who previously qualified for a tax exemption certificate, and registration for payment under deduction, applicable to all other subcontractors. The three-year qualifying period for gross payment will be reduced to one year (but the Revenue may cancel registration at any time where the qualifying conditions no longer apply, or where the rules have been breached). Unlike the present scheme contractors will be able to pay un-registered sub-contractors, but the deduction rate will be much higher (probably around 30%).Those who hold a subcontractor's certificate when the new provisions come into effect will be treated as being registered for gross payment and those who hold a registration card will be treated as registered for payment for deduction. Subcontractors holding temporary registration that expire before 6 April 2007 will have to register in the same way as new applicants.

For new workers, contractors will need to obtain basic identity details and will check with the Revenue what type of registration the worker holds. The verification may be done by telephone or over the internet. Contractors may assume that the status remains unchanged unless the Revenue notifies them to the contrary.

When contractors deduct tax from payments, they must supply the subcontractors with pay statements. Contractors will no longer be required to submit payment vouchers monthly. Instead, they will be required to submit monthly returns (even for those who make payments quarterly) and must provide details of recipients and payments made, together with a 'status declaration' that none of the payments relate to a contract of employment . This requirement puts the onus of establishing the worker's status on the contractor and there will be penalties for false declarations (and also various other penalties, as now, for both contractors and workers). The returns must be sent in every month, even if no payments have been made, and penalties will apply if a return is not submitted. Nil returns will be able to be made on paper, over the internet or by telephone. It is intended that other online services, such as subcontractor verification checks, will also be available.

## VAT

The general rule about liability to register for VAT is given in the VAT office notes. It is possible to give here only a brief outline of how the tax works. The rules that apply to the construction industry are extremely complex and all traders must study *The VAT Guide* and other publications.

Registration for VAT is required if, at the end of any month, the value of taxable supplies in the last 12 months exceeds the annual threshold or if there are reasonable grounds for believing the value of the taxable supplies in the next 30 days will exceed the annual threshold.

Taxable supplies include any zero-rated items. The annual threshold is £64,000. The amount of tax to be paid is the difference between the VAT charged out to customers (*output tax*) and that suffered on payments made to suppliers for goods and services (*input tax*) incurred in making taxable supplies. Unlike income tax there is no distinction in VAT for capital items so that the tax charged on the purchase of, for example, machinery, trucks and office furniture, will normally be reclaimable as *input tax*.

VAT is payable in respect of three monthly periods known as 'tax periods'. You can apply to have the group of tax periods that fits in best with your financial year. The tax must be paid within one month of the end of each tax period. Traders who receive regular repayments of VAT can apply to have them monthly rather than quarterly. Not all types of goods and services are taxed at 17.5% (i.e. the standard rate). Some are exempt and others are zero-rated.

### Zero-rated

This means that no VAT is chargeable on the goods or services, but a registered trader can reclaim any *input* tax suffered on his purchases. For instance, a builder pays VAT on the materials he buys to provide supplies of constructing but if he is constructing a new dwelling house, this is zero rated. The builder may reclaim this VAT or set it off against any VAT due on standard rated work.

### Exempt

Supplies that are exempt are less favourably treated than those that are zero rated. Again no VAT is chargeable on the goods or services but the trader cannot reclaim any *input* tax suffered on his purchases.

### Standard-rated

All work which is not specifically stated to be zero rated or exempt is standard-rated, i.e. VAT is chargeable at the current rate of 17.5% and the trader may deduct any *input* tax suffered when he is making his return to the Customs and Excise. If for any reason a trader makes a supply and fails to charge VAT when he should have done so (e.g. mistakenly assuming the supply to be zero rated), he will have to account for the VAT himself out of the proceeds. If there is any doubt about the VAT position, it is safer to assume the supply is standard rated, charge the appropriate amount of VAT on the invoice and argue about it later.

### Time of supply

The *time* at which a supply of goods or services is treated as taking place is important and is called the 'tax point'. VAT must be accounted for to the Customs and Excise at the end of the accounting period in which this 'tax point' occurs. For the supply of goods which are 'built on site', the 'basic tax point' is the date the goods are made available for the customer's use, whilst for *services* it is normally the date when all work except invoicing is completed.

However, if you issue a tax invoice or receive a payment before this 'basic tax point' then that date becomes a tax point. In the case of contracts providing for stage and retention payments, the tax point is either the date the tax invoice is issued or when payment is received, whichever is the earlier.

All the requirements apply to sub-contractors and main contractors and it should be noted that, when a contractor deducts income tax from a payment to a sub-contractor (because he has no valid CIS6) VAT is payable on the full gross amount *before* taking off the income tax.

### Annual accounting

It is possible to account for VAT other than on a specified three month period. Annual accounting provides for nine equal installments to be paid by direct debit with annual return provided with the tenth payment. £300,000.

### Cash accounting

If turnover is below a specified limit, currently £660,000, a taxpayer may account for VAT on the basis of cash paid and received. The main advantages are automatic bad debt relief and a deferral of VAT payment where extended credit is given.

### Bad debts

Relief is available for debts over 6 months.

# Part Seven

## GENERAL CONSTRUCTION DATA

# GENERAL CONSTRUCTION DATA

## The metric system

Linear

| | | |
|---|---|---|
| 1 centimetre (cm) | = | 10 millimetres (mm) |
| 1 decimetre (dm) | = | 10 centimetres (cm) |
| 1 metre (m) | = | 10 decimetres (dm) |
| 1 kilometre (km) | = | 1000 metres (m) |

Area

| | | |
|---|---|---|
| 100 sq millimetres | = | 1 sq centimetre |
| 100 sq centimetres | = | 1 sq decimetre |
| 100 sq decimetres | = | 1 sq metre |
| 1000 sq metres | = | 1 hectare |

Capacity

| | | |
|---|---|---|
| 1 millilitre (ml) | = | 1 cubic centimetre (cm3) |
| 1 centilitre (cl) | = | 10 millilitres (ml) |
| 1 decilitre (dl) | = | 10 centilitres (cl) |
| 1 litre (l) | = | 10 decilitres (dl) |

Weight

| | | |
|---|---|---|
| 1 centigram (cg) | = | 10 milligrams (mg) |
| 1 decigram (dg) | = | 10 centigrams (mcg) |
| 1 gram (g) | = | 10 decigrams (dg) |
| 1 decagram (dag) | = | 10 grams (g) |
| 1 hectogram (hg) | = | 10 decagrams (dag) |

## Conversion equivalents (imperial/metric)

Length

| | | |
|---|---|---|
| 1 inch | = | 25.4 mm |
| 1 foot | = | 304.8 mm |
| 1 yard | = | 914.4 mm |
| 1 yard | = | 0.9144 m |
| 1 mile | = | 1609.34 m |

Area

| | | |
|---|---|---|
| 1 sq inch | = | 645.16 sq mm |
| 1 sq ft | = | 0.092903 sq m |
| 1 sq yard | = | 0.8361 sq m |
| 1 acre | = | 4840 sq yards |
| 1 acre | = | 2.471 hectares |

Liquid

| | | |
|---|---|---|
| 1 lb water | = | 0.454 litres |
| 1 pint | = | 0.568 litres |
| 1 gallon | = | 4.546 litres |

Horse-power

| | | |
|---|---|---|
| 1 hp | = | 746 watts |
| 1 hp | = | 0.746 kW |
| 1 hp | = | 33,000 ft.lb/min |

Weight

| | | |
|---|---|---|
| 1 lb | = | 0.4536 kg |
| 1 cwt | = | 50.8 kg |
| 1 ton | = | 1016.1 kg |

**Conversion equivalents (metric/imperial)**

Length

| | | |
|---|---|---|
| 1 mm | = | 0.03937 inches |
| 1 centimetre | = | 0.3937 inches |
| 1 metre | = | 1.094 yards |
| 1 metre | = | 3.282 ft |
| 1 kilometre | = | 0.621373 miles |

Area

| | | |
|---|---|---|
| 1 sq millimetre | = | 0.00155 sq in |
| 1 sq metre | = | 10.764 sq ft |
| 1 sq metre | = | 1.196 sq yards |
| 1 acre | = | 4046.86 sq m |
| 1 hectare | = | 0.404686 acres |

Weight

| | | |
|---|---|---|
| 1 kg | = | 2.205 lbs |
| 1 kg | = | 0.01968 cwt |
| 1 kg | = | 0.000984 ton |

**Temperature equivalents**

In order to convert Fahrenheit to Celsius deduct 32 and multiply by 5/9.
To convert Celsius to Fahrenheit multiply by 9/5 and add 32.

| Fahrenheit | Celsius |
|---|---|
| 230 | 110.0 |
| 220 | 104.4 |
| 210 | 98.9 |
| 200 | 93.3 |
| 190 | 87.8 |
| 180 | 82.2 |
| 170 | 76.7 |
| 160 | 71.1 |
| 150 | 65.6 |
| 140 | 60.0 |
| 130 | 54.4 |
| 120 | 48.9 |
| 110 | 43.3 |
| 100 | 37.8 |
| 90 | 32.2 |
| 80 | 26.7 |
| 70 | 21.1 |
| 60 | 15.6 |
| 50 | 10.0 |
| 40 | 4.4 |
| 30 | -1.1 |
| 20 | -6.7 |
| 10 | -12.2 |
| 0 | -17.8 |

## Areas and volumes

| Figure | Area | Perimeter |
|---|---|---|
| Rectangle | Length × breadth | Sum of sides |
| Triangle | Base × half of perpendicular height | Sum of sides |
| Quadrilateral | Sum of areas of contained triangles | Sum of sides |
| Trapezoidal | Sum of areas of contained triangles | Sum of sides |
| Trapezium | Half of sum of parallel sides × perpendicular height | Sum of sides |
| Parallelogram | Base × perpendicular height | Sum of sides |
| Regular polygon | Half sum of sides × half internal diameter | Sum of sides |
| Circle | pi × radius$^2$ | pi × diameter or pi × 2 × radius |

| Figure | Surface area | Volume |
|---|---|---|
| Cylinder | pi × 2 × radius$^2$ × length (curved surface only) | pi × 2 × radius$^2$ × length |
| Sphere | pi × diameter$^2$ | Diameter$^3$ × 0.5236 |

| Weights of materials | kg/m3 | kg/m$^2$ | kg/m |
|---|---|---|---|
| Aggregate, coarse | 1,500 | | |
| Ashes | 800 | | |
| Ballast | 600 | | |

| **Weights of materials** | **kg/m3** | **kg/m²** | **kg/m** |
|---|---|---|---|
| Blocks, natural aggregate | | | |
| 75mm | | 160 | |
| 100mm | | 215 | |
| 140mm | | 300 | |
| Blocks, lighweight aggregate | | | |
| 75mm | | 60 | |
| 100mm | | 80 | |
| 140mm | | 112 | |
| Bricks, Fletton | | 1820 | |
| Bricks, engineering | | 2250 | |
| Bricks, concrete | | 1850 | |
| Brickwork, 112.5mm | | 220 | |
| Brickwork, 215mm | | 465 | |
| Brickwork, 327.5mm | | 710 | |
| Cement | 1,440 | | |
| Chalk | 2,240 | | |
| Flint | 2,550 | | |
| Gravel | 1,750 | | |
| Hardcore | 1,900 | | |
| Hoggin | 1,750 | | |
| Lime, ground | 750 | | |
| Stone, natural | 2,400 | | |
| Stone, Portland | 2,200 | | |
| Stone, reconstructed | 2,250 | | |
| Stone, York | 2,400 | | |

## EXCAVATION AND FILLING

### Shrinkage of deposited material

| | |
|---|---|
| Clay | -10% |
| Gravel | -7.50% |
| Sandy soil | -12.50% |

## Bulking excavated material

| | |
|---|---|
| Clay | 40% |
| Gravel | 25% |
| Sand | 20% |

## Typical fuel consumption for plant

| | Engine size kW | Litres per hour |
|---|---|---|
| Compressors up to | 20.00 | 4.00 |
| | 30.00 | 6.50 |
| | 40.00 | 8.20 |
| | 50.00 | 9.00 |
| | 75.00 | 16.00 |
| | 100.00 | 20.00 |
| | 125.00 | 25.00 |
| | 150.00 | 30.00 |
| Concrete mixers up to | 5.00 | 1.00 |
| | 10.00 | 2.40 |
| | 15.00 | 3.80 |
| | 20.00 | 5.00 |
| | 20.00 | 5.00 |
| Dumpers | 5.00 | 1.30 |
| | 7.00 | 2.00 |
| | 10.00 | 3.00 |
| | 15.00 | 4.00 |
| | 20.00 | 4.90 |
| | 30.00 | 7.00 |
| | 50.00 | 12.00 |
| Excavators | 10.00 | 2.50 |
| | 20.00 | 4.50 |
| | 40.00 | 9.00 |
| | 60.00 | 13.00 |
| | 80.00 | 17.00 |

| | Engine size kW | Litres per hour |
|---|---|---|
| Pumps | 5.00 | 1.10 |
| | 10.00 | 2.10 |
| | 15.00 | 3.20 |
| | 20.00 | 4.20 |
| | 25.00 | 5.50 |

## CONCRETE WORK

| Concrete mixes | Cement t | Sand m3 | Aggregate m3 | Water litres |
|---|---|---|---|---|
| 1:1:2 | 0.50 | 0.45 | 0.70 | 208.00 |
| 1:1:5:3 | 0.37 | 0.50 | 0.80 | 185.00 |
| 1:2:4 | 0.30 | 0.54 | 0.85 | 175.00 |
| 1:3:6 | 0.22 | 0.55 | 0.85 | 160.00 |

## BRICKWORK AND BLOCKWORK

| Bricks per m2 (brick size 215 × 103.5 × 65mm) | nr |
|---|---|
| Half brick wall | |
| stretcher bond | 59 |
| English bond | 89 |
| English garden wall bond | 74 |
| Flemish bond | 79 |
| One brick wall | |
| English bond | 118 |
| Flemish bond | 118 |

| **Bricks per m2 (cont'd)** | **nr** |
|---|---|
| One and a half brick wall | |
| English bond | 178 |
| Flemish bond | 178 |
| Two brick wall | |
| English bond | 238 |
| Flemish bond | 238 |
| Metric modular bricks | |
| 200 × 100 × 75mm | |
| 90mm thick | 133 |
| 190mm thick | 200 |
| 200 × 100 × 100mm | |
| 90mm thick | 50 |
| 190mm thick | 100 |
| 290mm thick | 150 |
| 300 × 100 × 75mm | |
| 90mm thick | 44 |
| 300 × 100 × 100mm | |
| 90mm thick | 50 |

Blocks per m2 (block size 414 × 215mm)

| | |
|---|---|
| 60mm thick | 9.9 |
| 75mm thick | 9.9 |
| 100mm thick | 9.9 |
| 140mm thick | 9.9 |
| 190mm thick | 9.9 |
| 215mm thick | 9.9 |

| Mortar per m2 | Wirecut m3 | 1 Frog m3 | 2 Frogs m3 |
|---|---|---|---|
| Brick size 215 × 103.5 × 65mm | | | |
| Half brick wall | 0.017 | 0.024 | 0.031 |
| One brick wall | 0.045 | 0.059 | 0.073 |
| One and a half brick wall | 0.072 | 0.093 | 0.114 |
| Two brick wall | 0.101 | 0.128 | 0.155 |

| Brick size 200 × 100 × 75mm | Solid m3 | Perforated m3 |
|---|---|---|
| 90mm thick | 0.016 | 0.019 |
| 190mm thick | 0.042 | 0.048 |
| 290mm thick | 0.068 | 0.078 |
| **Brick size 200 × 100 × 100mm** | | |
| 90mm thick | 0.013 | 0.016 |
| 190mm thick | 0.036 | 0.041 |
| 290mm thick | 0.059 | 0.067 |
| **Brick size 200 × 100 × 100mm** | | |
| 90mm thick | 0.015 | 0.018 |
| **Block size 440 × 215mm** | | |
| 60mm thick | 0.004 | |
| 75mm thick | 0.005 | |
| 100mm thick | 0.006 | |
| 140mm thick | 0.007 | |
| 190mm thick | 0.008 | |
| 215mm thick | 0.009 | |

## MASONRY

| Mortar per m2 of random rubble walling | m3 |
|---|---|
| 300mm thick wall | 0.120 |
| 450mm thick wall | 0.160 |
| 550mm thick wall | 0.120 |

## EXTERNAL WORKS

| Blocks/slabs per m2 | nr/m2 |
|---|---|
| 200 × 100mm | 50.00 |
| 450 × 450mm | 4.93 |
| 600 × 450mm | 3.70 |
| 600 × 600mm | 2.79 |
| 600 × 750mm | 2.22 |
| 600 × 900mm | 1.85 |

| Drainage trench widths | Under 1.5m deep mm | Over 1.5m deep mm |
|---|---|---|
| Pipe diameter 100mm | 450 | 600 |
| Pipe diameter 150mm | 500 | 650 |
| Pipe diameter 225mm | 600 | 750 |
| Pipe diameter 300mm | 650 | 800 |

| Volumes of filling for pipe beds (m3 per m) | 50mm thick | 100mm thick | 150mm thick |
|---|---|---|---|
| Pipe diameter 100mm | 0.023 | 0.045 | 0.068 |
| Pipe diameter 150mm | 0.026 | 0.053 | 0.079 |
| Pipe diameter 225mm | 0.030 | 0.060 | 0.090 |
| Pipe diameter 300mm | 0.038 | 0.075 | 0.113 |

| **Volumes of filling for pipe bed and haunching (m3 per m)** | **m3** |
|---|---|
| Pipe diameter 100mm | 0.117 |
| Pipe diameter 150mm | 0.152 |
| Pipe diameter 225mm | 0.195 |
| Pipe diameter 300mm | 0.279 |

| **Volumes of filling for pipe bed and surround (m3 per m)** | **m3** |
|---|---|
| Pipe diameter 100mm | 0.185 |
| Pipe diameter 150mm | 0.231 |
| Pipe diameter 225mm | 0.285 |
| Pipe diameter 300mm | 0.391 |

# Part Eight

## PLANT NAMES

## ENGLISH/LATIN PLANT NAMES

| Common/English | Scientific/Latin |
|---|---|
| Adam's Needle | Yucca gloriosa |
| Alder | Alnus |
| Alder Buckhorn | Frangula alnus |
| Algerian Iris | Iris unguicularis |
| Apple | Malus |
| Ash | Fraxinus |
| Atlantic Cedar | Cedrus libani atlantica |
| Austrian Pine | Pinus nigra austriaca |
| Balsam Poplar | Populus balsamifera |
| Bamboo | Pseudosasa japonica |
| Bay | Laurus nobilis |
| Bay Willow | Salix pentandra |
| Bear Tree | Arctostaphylos uva-ursi |
| Beech | Fagus |
| Birch | Betula |
| Bird Cherry | Prunus padus |
| Black Poplar | Populus nigra |
| Black Stemmed Bamboo | Phyllostachys nigra |
| Black Walnut | Juglans nigra |
| Blackthorn | Prunus spinosa |
| Bladder Nut | Staphylea colchica |
| Blue Holly | Ilex x meserveae |
| Boston Ivy | Parthenocissus tricuspidata |
| Box | Buxus |
| Bramble | Rubus |
| Broom | Cytisus |
| Broom | Cytisus scorparius |
| Bugle | Ajuga |
| Butchers Broom | Ruscus aculeatus |
| Butterbur | Petastites hybidus |
| Canary Ivy | Hedera canarienis |
| Cape Figwort | Phygelius capensis coccineus |
| Cedar | Cedrus |

| Common/English | Scientific/Latin |
|---|---|
| Cherry | Prunus |
| Cherry Laurel | Prunus laurocreasus |
| Chestnut Holly | Ilex x koeheana |
| Christmas Box | Sarcococca |
| Chusan Palm | Trachycarpus fortunei |
| Comfrey | Symphytum |
| Common Alder | Alnus glutinosa |
| Common Ivy | Hedera helix |
| Corsican Pine | Pinus nigra maritima |
| Crab Apple | Malus |
| Crack Willow | Salix fragilis |
| Cranesbill | Geranium |
| Cricket Bat Willow | Salix alba Caerulea |
| Cucumber Tree | Magnolia acuminata |
| Current | Ribes |
| Cypress | Cupressus |
| Daisy Bush | Olearia |
| Dawn Redwood | Metasequoia glyptostroboides |
| Day Lily | Hemerocallis |
| Dog Rose | Rosa canina |
| Dogwood | Cornus |
| Douglas Fir | Pseudotsuga menziesii |
| Downy Birch | Betula pubescens |
| Dwarf Almond | Prunus tenella |
| Dyers Greenweed | Genista tinctoria |
| Eglantine or Sweet Briar | Rosa rubiginosa |
| Elder | Sambucus |
| Elephant's Ears | Bergenia |
| Elisha's Tears | Leycesreria Formosa |
| Evergreen Magnolia | Magnolia grandiflora |
| Evergreen Oak | Quercus ilex |
| False Acacia | Robinia pseudoacacia |
| Field Maple | Acer campestre |

| **Common/English** | **Scientific/Latin** |
|---|---|
| Field Rose | Rosa arvenis |
| Fig | Ficus carica |
| Fir | Abies |
| Firethorn | Pyracantha |
| Flowering Ash | Fraxinus ornus |
| Flowering Quince | Chaenomeles |
| Fly Honeysuckle | Lonicera xylostenum |
| Foxglove Tree | Paulownia |
| | |
| Germander | Teucrium |
| Giants Reed | Arundo donax |
| Giant Rhubarb | Gunnera manicata |
| Gladwyn | Iris foetidissima |
| Goat Willow/Sallow | Salix caprea |
| Goats Beard | Aruncus dioicus |
| Golden Hop | Humulus lupulus aureus |
| Golden Oat | Stipa gigantean |
| Golden Rain | Koelreuteria paniculata |
| Gorse | Ulex |
| Grapevine | Vitis vinifera |
| Grey Alder | Alnus incana |
| Guelder Rose | Viburnum opulus |
| | |
| Handkerchief Tree | Davidia involucrate |
| Hard Fern | Blechnum spicant |
| Hawthorn | Crataegus monogyna |
| Hazel | Corylus avellana |
| Heath | Erica |
| Heather | Calluna vulgaris |
| Hedging Privet | Ligustrum ovalifolium |
| Hemlock | Tsuga heterophylla |
| Holly | Ilex |
| Holly | Ilex aquifolium |
| Honey Locust | Gleditsia |
| Honeysuckle | Lonicera |
| Hop Hornbeam | Ostrya virginina |

| Common/English | Scientific/Latin |
|---|---|
| Hop Tree | Ptelea trifoliate |
| Hornbeam | Carpinus betulus |
| Horse Chestnut | Aesculus hippocastanum |
| Hungarian Oak | Quercus frainetto |
| Hyssop | Hyssopus officinalis |
| | |
| Indian Bean Tree | Catalpa bignonioides |
| Indian Chestnut | Aesculus indica |
| Irish Yew | Taxus baccata Fastigiata |
| Iron Tree | Parrotia persica |
| Italian Alder | Alnus cordata |
| Italian Cypress | Cupressus sempervirens Stricta |
| Ivy | Hedera |
| | |
| Japanese Anemone | Anemone hupehensis japonic |
| Japanese Anemone | Anemone x hybrida |
| Japanese Cedar | Cryptomeria japonica |
| Japanese Honeysuckle | Lonicera japonica |
| Japanese Larch | Larix kaempferi |
| Japanese Maple | Acer palmatum |
| Jasmine | Jasminium officinale |
| Jerusalem Sage | Phlomis fruiticosa |
| Judas Tree | Cercis siliquastrum |
| Juniper | Juniperus communis |
| | |
| Kiwi Fruit | Actinida chinensis |
| Knotweed | Polygonum |
| | |
| Lads Love | Artemisia abrotanum |
| Lady Fern | Athyrium fillix-femina |
| Larch | Larix |
| Large-Leaved Lime | Tilia platyphyllos |
| Lauristinus | Viburnum tinus |
| Lavender | Lavandula |
| Lavender Cotton | Santolina |
| Lawson's Cypress | Chamaecyparis lawsoniana |

| **Common/English** | **Scientific/Latin** |
|---|---|
| Lebanese Cedar | Cedrus libani |
| Lenten Rose | Hellborus orientails |
| Leyland Cypress | Cupressocyparis leylandii |
| Lilac | Syringa |
| Lily Turf | Liriope muscari |
| Lime | Tilia |
| London Plane | Platanus acerifolia |
| Loquat | Eriobotrya japonica |
| | |
| Male Fern | Dryopteris filix-mas |
| Maple | Acer |
| Medlar | Mespilus germanica |
| Mexican Orange | Choisya |
| Mimosa | Acacia dealbata |
| Mock Orange | Philadelphus |
| Morello Cherry | Prunus avium Morello |
| Moss Rose | Rosa x centifolia Muscosa |
| Mulberry | Morus nigra |
| Myrtle | Myrtus communis |
| | |
| New Zealand Flax | Phormium |
| Norway Maple | Acer platanoides |
| Norway Spruce | Picea abies |
| | |
| Oak | Quercus robur |
| Oregon Grape | Mahonia aquifolium |
| Osier | Salix viminalis |
| | |
| Pagoda Tree | Sophora japonica |
| Pampas Grass | Cortaderia |
| Passion Flower | Passiflora caerulea |
| Pear | Pyrus |
| Periwinkle | Vinca |
| Persian Ivy | Hedera colchica |
| Pine | Pinus |
| Plane | Platanus |

| Common/English | Scientific/Latin |
|---|---|
| Polypody | Polypodium |
| Poplar | Populus |
| Portuguese Laurel | Prunus lusitanica |
| Privet | Ligustrum |
| | |
| Quince | Cydonia oblonga |
| | |
| Red Hot Poker | Kniphofla |
| Red Oak | Quercus rubra |
| Redwood | Sequoia sempervirens |
| Rock Rose | Helianthemum |
| Rosa | Rose |
| Rose of Sharon | Hypericum calcyinum |
| Rosemary | Rosmarinus |
| Rowan | Sorbus aucuparia |
| Royal Fern | Osmunda regalis |
| Rue | Ruta graveoleans |
| | |
| Sage | Salvia |
| Sallow | Salix cinerea |
| Salt Bush | Atriplex halimus |
| Scarlet Oak | Quercus coccinea |
| Scotch Rose | Rosa pimpinellilfolia |
| Scots Pine | Pinus sylvestris |
| Sea Buckhorn | Hippophae rhamnoides |
| Sedge | Carex |
| Shrubby Cinqfoil | Potentilla |
| Shuttlecock Fern | Matteuccia struthiopteris |
| Silver Birch | Betula pendula |
| Silver Lime | Tilia tomentosa |
| Silver Maple | Acer saccharinum |
| Small-Leaved Lime | Tilia cordata |
| Smoke Bush | Cotinus |
| Snowberry | Symphoricarpus |
| Snowy Mespilus | Amelanchier |
| Southern Beech | Nothofagus |

| **Common/English** | **Scientific/Latin** |
|---|---|
| Spanish Broom | Spartium junceum |
| Spanish Chestnut | Castanea sativa |
| Spindle | Spindle Euonymus |
| Spindle Tree | Euonymus europaeus |
| Spuce | Picea |
| Spurge | Euphorbia |
| Spurge Laurel | Rosa rubiginosa |
| Sycamore | Acer pseudoplatanus |
| | |
| Tamarisk | Tamarix |
| Tealeaf Willow | Salix phylicifolia |
| Thyme | Thymus |
| Tree Lupin | Lupinus arboreus |
| Tree Peony | Paeonia |
| Tree Poppy | Romneya coulteri |
| Tree Heaven | Ailanthus |
| Tulip Tree | Liriodedrun tulipifera |
| Turkey Oak | Quercus cerris |
| | |
| Vine | Vitis |
| Virginia Creeper | Parthenocissus quinquefolia |
| Violet Willow | Salix daphnoides |
| | |
| Walnut | Juglans regia |
| Wayfaring Tree | Viburnum lantana |
| Weeping Willow | Salix sepuilchralis Chrysocoma |
| Wellingtonia | Sequoiadendron giganteum |
| Western Red Cedar | Thuja plicata |
| White Mulberry | Morus alba |
| White Poplar | Populus alba |
| White Willow | Salix alba |
| Whitebeam | Sorbus aria |
| Wild Apple | Malus sylvestris |
| Wild Privet | Ligustrum vulgare |
| Wild Service Tree | Sorbus torminalis |
| Willow/Sallow | Salix |

| Common/English | Scientific/Latin |
|---|---|
| Willow | Salix reifenweide |
| Wing Nut | Pterocaya fraxinifolia |
| Winter Honeysuckle | Lonicera x purpusii |
| Winter Jasmine | Jasminium nudiflorum |
| Winterswee | Chimonanthust |
| Witch Hazel | Hamamelis |
| Wood Spurge | Euphorbia amygdaloides |
| Woodland Hawthorn | Crataegus oxyacantha |
| Woody Nightshade | Solanum dulcamara |
| Woody Willow | Salix lanata |
| Wormwood | Artemisia absinthium |
| Yarrow | Achillea millefolium |
| Yew | Taxus |

## LATIN/ENGLISH PLANT NAMES

| Scientific/Latin | Common/English |
|---|---|
| Abies | Fir |
| Acacia dealbata | Mimosa |
| Acer | Maple |
| Acer campestre | Field Maple |
| Acer palmatum | Japanese Maple |
| Acer platanoides | Norway Maple |
| Acer pseudoplatanus | Sycamore |
| Acer saccharinum | Silver Maple |
| Acer saccharum | Sugar Maple |
| Achillea millefolium | Yarrow |
| Actinida chinensis | Kiwi Fruit |
| Aesculus hippocastanum | Horse Chestnut |
| Aesculus indica | Indian Chestnut |
| Ailanthus alitissima | Tree of Heaven |
| Ajuga | Bugle |
| Alnus | Alder |

| Scientific/Latin | Common/English |
|---|---|
| Alnus cordata | Italian Alder |
| Alnus glutinosa | Common Alder |
| Alnus incana | Grey Alder |
| Amelanchier | Snowy Mespilus |
| Anemone hupehensis japonica | Japanese Anemone |
| Anemone x hybrida | Japanese Anemone |
| Arbutus unedo | Strawberry Tree |
| Arctostaphylos uva-ursi | Bear Tree |
| Artemisia abrotanum | Lads Love |
| Artemisia absinthium | Wormwood |
| Aruncus dioicus | Goats Beard |
| Arundo donax | Giants Reed |
| Athyrium fillix-femina | Lady Fern |
| Atriplex halimus | Salt Bush |
| | |
| Bergenia | Elephant's Ears |
| Betula | Birch |
| Betula papyrifera | Paper Birch |
| Betula pendula | Silver Birch |
| Betula pubescens | Downy Birch |
| Blechnum spicant | Hard Fern |
| Buxus | Box |
| Buxus sempervirens | Box |
| | |
| Calluna vulgaris | Heather |
| Carex | Sedge |
| Carpinus betulus | Hornbeam |
| Castanea sativa | Spanish Chestnut |
| Catalpa bignonioides | Indian Bean Tree |
| Cedrus | Cedar |
| Cedrus libani | Lebanese Cedar |
| Cedrus libani atlantica | Atlantic Cedar |
| Cercis siliquastrum | Judas Tree |
| Chaenomeles | Flowering Quince |
| Chamaecyparis lawsoniana | Lawson's Cypress |
| Chimonanthus | Wintersweet |

| Scientific/Latin | Common/English |
|---|---|
| Choisya | Mexican Orange |
| Cistus | Sun Rose |
| Cornus | Dogwood |
| Cortaderia | Pampas Grass |
| Corylus avellana | Hazel |
| Cotinus | Smoke Bush |
| Crataegus monogyna | Hawthorn |
| Crataegus oxyancantha | Woodland Hawthorn |
| Cryptomeria japonica | Japanese Cedar |
| Cupressocyparis leylandii | Leyland Cypress |
| Cupressus | Cypress |
| Cupressus sempervirens Stricta | Italian Cypress |
| Cydonia oblonga | Quince |
| Cytisus | Broom |
| Cytisus scorparius | Broom |
| Daphne laureola | Spurge Laurel |
| Davidia involucrata | Handkerchief Tree |
| Dryopteris filix-mas | Male Fern |
| Erica | Heath |
| Eriobotrya japonica | Loquat |
| Euonymus | Spindle |
| Euonymus europaeus | Spindle Tree |
| Euphorbia | Spurge |
| Euphorbia amygadaloides | Wood Spurge |
| Fagus | Beech |
| Ficus carica | Fig |
| Frangula alnus | Alder Buckhorn |
| Fraxinus | Ash |
| Fraxinus ornus | Flowering Ash |
| Genista tinctoria | Dyers Greenweed |
| Geranium | Cranesbill |
| Gleditsia | Honey Locust |

| **Scientific/Latin** | **Common/English** |
|---|---|
| Gunnera manicata | Giant Rhubarb |
| | |
| Hamamelis | Witch Haze |
| Hedera | Ivy |
| Hedera canarienis | Canary Ivy |
| Hedera colchica | Persian Ivy |
| Hedera helix | Ivy |
| Helianthemum | Rock Rose |
| Hellborus foetidus | Stinking Hellebore |
| Hellborus orientalis | Lenten Rose |
| Hemerocallis | Day Lily |
| Hippophae rhamnoides | Sea Buckhorn |
| Humulus lupulus aureus | Golden Hop |
| Hypericum calcyinum | Rose of Sharon |
| Hyssopus officinalis | Hyssop |
| | |
| Ilex | Holly |
| Ilex aquifolium | Holly |
| Ilex x koeheana | Chestnut Holly |
| Ilex x meserveae | Blue Holly |
| Iris foetidissima | Gladwyn |
| Iris unguicularis | Algerian Iris |
| | |
| Jasminium nudiflorum | Winter Jasmine |
| Jasminium officinale | Jasmine |
| Juglans nigra | Black Walnut |
| Juglans regia | Walnut |
| Juniperus | Juniper |
| Juniperus communis | Juniper |
| | |
| Kniphofla | Red Hot Poker |
| Koelreute | Golden Rain |
| | |
| Larix | Larch |
| Larix kaempferi | Japanese Larch |
| Laurus nobilis | Bay |

| Scientific/Latin | Common/English |
|---|---|
| Lavandula | Lavender |
| Leycesreria formosa | Elisha's Tears |
| Ligustrum | Privet |
| Ligustrum ovalifolium | Hedging Privet |
| Ligustrum vulgare | Wild Privet |
| Liriodedrun tulipifera | Tulip Tree |
| Liriope muscari | Lily Turf |
| Lonicera | Honeysuckle |
| Lonicera japonica | Japanese Honeysuckle |
| Lonicera x purpusii | Winter Honeysuckle |
| Lonicera xylostenum | Fly Honeysuckle |
| Lupinus arboreus | Tree Lupin |
| | |
| Magnolia acuminata | Cucumber Tree |
| Magnolia grandiflora | Evergreen Magnolia |
| Mahonia aquifolium | Oregon Grape |
| Malus | Apple |
| Malus sylvestris | Wild Apple |
| Matteuccia struthiopteris | Shuttlecock Fern |
| Mespilus germanica | Medlar |
| Metasequoia glyptostroboides | Dawn Redwood |
| Morus alba | White Mulberry |
| Morus nigra | Mulberry |
| Myrtus communis | Myrtle |
| | |
| Nothofagus | Southern Beech |
| | |
| Olearia | Daisy Bush |
| Osmunda regalis | Royal Fern |
| Ostrya virgini | Hop Hornbeam |
| | |
| Paeonia | Tree Peony |
| Parrotia persica | Iron Tree |
| Parthenocissus quinquefolia | Virgina Creeper |
| Parthenocissus tricuspidata | Boston Ivy |
| Passiflora caerulea | Passion Flower |

| **Scientific/Latin** | **Common/English** |
|---|---|
| Paulownia | Foxglove Tree |
| Petastites hybidus | Butterbur |
| Philadelphus | Mock Orange |
| Phlomis fruiticosa | Jerusalem Sage |
| Phormium | New Zealand Flax |
| Phygelius capensis coccineus | Cape Figwort |
| Phyllostachys nigra | Black Stemmed Bamboo |
| Picea | Spruce |
| Picea abies | Norway Spruce |
| Picea omorika | Serbian Spruce |
| Picea sitchensis | Sitka Spruce |
| Pinus | Pine |
| Pinus nigra austriaca | Austrian Pine |
| Pinus nigra maritime | Corsican Pine |
| Pinus sylvestris | Scots Pine |
| Platanus | Plane |
| Platanus acerifolia | London Plane |
| Polygonum | Knotweed |
| Polypodium | Polypody |
| Populus | Poplar |
| Populus alba | White Poplar |
| Populus balsamifera | Balsam Poplar |
| Populus nigra | Black Poplar |
| Potentilla | Shrubby Cinqfoil |
| Prunus | Cherry |
| Prunus avium Morello | Morello Cherry |
| Prunus laurocreasus | Cherry Laurel |
| Prunus lusitanica | Portuguese Laurel |
| Prunus mahaleb | St Lucia Cherry |
| Prunus padus | Bird Cherry |
| Prunus spinosa | Blackthorn |
| Prunus tenella | Dwarf Almond |
| Pseudosasa japonica | Bamboo |
| Pseudotsuga menziesii | Douglas Fir |
| Ptelea trifoliata | Hop Tree |
| Pterocarya fraxinifolia | Wing Nut |

| Scientific/Latin | Common/English |
|---|---|
| Pyracantha | Firethorn |
| Pyrus | Pear |
| | |
| Quercus | Oak |
| Quercus cerris | Turkey Oak |
| Quercus coccinea | Scarlet Oak |
| Quercus frainetto | Hungarian Oak |
| Quercus ilex | Evergreen Oak |
| Quercus petraea | Sessile Oak |
| Quercus robur | Oak |
| Quercus rubra | Red Oak |
| | |
| Rhus typhinia | Sumach |
| Ribes | Current |
| Robinia pseudoacacia | False Acacia |
| Romneya coulteri | Tree Poppy |
| Rosa | Rose |
| Rosa arvenis | Field Rose |
| Rosa canina | Dog Rose |
| Rosa pimpinellilfolia | Scotch Rose |
| Rosa rubiginosa | Eglantine or Sweet Briar |
| Rosa x centifolia Muscosa | Moss Rose |
| Rosmarinus | Rosemary |
| Rubus | Bramble |
| Ruscus aculeatus | Butchers Broom |
| Ruta graveoleans | Rue |
| | |
| Salix | Willow/Sallow |
| Salix alba | White Willow |
| Salix alba Caerulea | Cricket Bat Willow |
| Salix caprea | Goat Willow/Sallow |
| Salix cinerea | Sallow |
| Salix daphnoides | Violet Willow |
| Salix fragilis | Crack Willow |
| Salix lanata | Woody Willow |
| Salix pentandra | Bay Willow |

| **Scientific/Latin** | **Common/English** |
|---|---|
| Salix phylicifolia | Tealeaf Willow |
| Salix viminalis | Osier |
| Salix x sepulchralis Chrysocoma | Weeping Willow |
| Salvia | Sage |
| Sambucus | Elder |
| Santolina | Lavender Cotton |
| Sarcococca | Christmas Box |
| Sedum | Stonecrop |
| Sequoia sempervirens | Redwood |
| Sequoiadendron giganteum | Wellingtonia |
| Solanum dulcamara | Woody Nightshade |
| Sophora japonica | Pagoda Tree |
| Sorbus aria | Whitebeam |
| Sorbus aucuparia | Rowan |
| Sorbus terminalis | Wild Service Tree |
| Spartium junceum | Spanish Broom |
| Staphylea colchica | Bladder Nut |
| Stipa gigantea | Golden Oat |
| Symphoricarpus | Snowberry |
| Symphytum | Comfrey |
| Syringa | Lilac |
| Tamarix | Tamarisk |
| Taxus | Yew |
| Taxus baccata Fastigiata | Irish Yew |
| Teucrium | Germander |
| Thuja plicata | Western Red Cedar |
| Thymus | Thyme |
| Tilia | Lime |
| Tilia cordata | Small-Leaved Lime |
| Tilia platyphyllos | Large-Leaved Lime |
| Tilia tomentosa | Silver Lime |
| Trachycarpus fortunei | Chusan Palm |
| Tsuga heterophylla | Hemlock |
| Ulex | Gorse |

| Scientific/Latin | Common/English |
|---|---|
| Viburnum lantana | Wayfaring Tree |
| Viburnum opulus | Guelder Rose |
| Viburnum tinus | Lauristinus |
| Vinca | Periwinkle |
| Vitis | Vine |
| Vitis vinifera | Grapevine |
| Yucca gloriosa | Adam's Needle |

# Index

Printed in Great Britain
by Amazon.co.uk, Ltd.,
Marston Gate.